Selected Titles in This Series

741 **Francisco Santos,** Triangulations of oriented matroids, 2002

740 **Rick Durrett,** Mutual invadability implies coexistence in spatial models, 2002

739 **Georgios K. Alexopoulos,** Sub-Laplacians with drift on Lie groups of polynomial volume growth, 2002

738 **Yasuro Gon,** Generalized Whittaker functions on $SU(2,2)$ with respect to the Siegel parabolic subgroup, 2002

737 **Arjen Doelman, Robert A. Gardner, and Tasso J. Kaper,** A stability index analysis of 1-D patterns of the Gray-Scott model, 2002

736 **Wojciech Chachólski and Jérôme Scherer,** Homotopy theory of diagrams, 2002

735 **Martina Brück, Xi Du, Joonsang Park, and Chuu-Lian Terng,** The submanifold geometries associated to Grassmannian systems, 2002

734 **Michel Van den Bergh,** Blowing up of non-commutative smooth surfaces, 2001

733 **Milé Krajčevski,** Tilings of the plane, hyperbolic groups and small cancellation conditions, 2001

732 **Jan O. Kleppe, Juan C. Migliore, Rosa Miró-Roig, Uwe Nagel, and Chris Peterson,** Gorenstein liaison, complete intersection liaison invariants and unobstructedness, 2001

731 **Jesús Bastero, Mario Milman, and Francisco J. Ruiz,** On the connection between weighted norm inequalities, commutators and real interpolation, 2001

730 **Suhyoung Choi,** The decomposition and classification of radiant affine 3-manifolds, 2001

729 **Michael Grosser, Eva Farkas, Michael Kunzinger, and Roland Steinbauer,** On the foundations of nonlinear generalized functions I and II, 2001

728 **Laura Smithies,** Equivariant analytic localization of group representations, 2001

727 **Anthony D. Blaom,** A geometric setting for Hamiltonian perturbation theory, 2001

726 **Victor L. Shapiro,** Singular quasilinearity and higher eigenvalues, 2001

725 **Jean-Pierre Rosay and Edgar Lee Stout,** Strong boundary values, analytic functionals, and nonlinear Paley-Wiener theory, 2001

724 **Lisa Carbone,** Non-uniform lattices on uniform trees, 2001

723 **Deborah M. King and John B. Strantzen,** Maximum entropy of cycles of even period, 2001

722 **Hernán Cendra, Jerrold E. Marsden, and Tudor S. Ratiu,** Lagrangian reduction by stages, 2001

721 **Ingrid C. Bauer,** Surfaces with $K^2 = 7$ and $p_g = 4$, 2001

720 **Palle E. T. Jorgensen,** Ruelle operators: Functions which are harmonic with respect to a transfer operator, 2001

719 **Steve Hofmann and John L. Lewis,** The Dirichlet problem for parabolic operators with singular drift terms, 2001

718 **Bernhard Lani-Wayda,** Wandering solutions of delay equations with sine-like feedback, 2001

717 **Ron Brown,** Frobenius groups and classical maximal orders, 2001

716 **John H. Palmieri,** Stable homotopy over the Steenrod algebra, 2001

715 **W. N. Everitt and L. Markus,** Multi-interval linear ordinary boundary value problems and complex symplectic algebra, 2001

714 **Earl Berkson, Jean Bourgain, and Aleksander Pełczynski,** Canonical Sobolev projections of weak type $(1,1)$, 2001

713 **Dorina Mitrea, Marius Mitrea, and Michael Taylor,** Layer potentials, the Hodge Laplacian, and global boundary problems in nonsmooth Riemannian manifolds, 2001

712 **Raúl E. Curto and Woo Young Lee,** Joint hyponormality of Toeplitz pairs, 2001

(Continued in the back of this publication)

Triangulations of Oriented Matroids

MEMOIRS
of the
American Mathematical Society

Number 741

Triangulations of Oriented Matroids

Francisco Santos

March 2002 • Volume 156 • Number 741 (second of 5 numbers) • ISSN 0065-9266

American Mathematical Society
Providence, Rhode Island

2000 *Mathematics Subject Classification.*
Primary 52C40; Secondary 52B11, 52B35.

Library of Congress Cataloging-in-Publication Data

Santos, Francisco, 1968–
Triangulations of oriented matroids / Francisco Santos.
p. cm. — (Memoirs of the American Mathematical Society, ISSN 0065-9266 ; no. 741)
"Volume 156, number 741 (second of 5 numbers)."
Includes bibliographical references.
ISBN 0-8218-2769-3
1. Oriented matroids. 2. Polytopes. 3. Triangularization (Mathematics) I. Title. II. Series.
QA3 .A57 no. 741
[QA166.6]
510 s—dc21
[511′.6] 2001056087

Memoirs of the American Mathematical Society

This journal is devoted entirely to research in pure and applied mathematics.

Subscription information. The 2002 subscription begins with volume 155 and consists of six mailings, each containing one or more numbers. Subscription prices for 2002 are $524 list, $419 institutional member. A late charge of 10% of the subscription price will be imposed on orders received from nonmembers after January 1 of the subscription year. Subscribers outside the United States and India must pay a postage surcharge of $31; subscribers in India must pay a postage surcharge of $43. Expedited delivery to destinations in North America $35; elsewhere $130. Each number may be ordered separately; *please specify number* when ordering an individual number. For prices and titles of recently released numbers, see the New Publications sections of the *Notices of the American Mathematical Society.*

Back number information. For back issues see the *AMS Catalog of Publications.*

Subscriptions and orders should be addressed to the American Mathematical Society, P. O. Box 845904, Boston, MA 02284-5904. *All orders must be accompanied by payment.* Other correspondence should be addressed to Box 6248, Providence, RI 02940-6248.

Memoirs of the American Mathematical Society is published bimonthly (each volume consisting usually of more than one number) by the American Mathematical Society at 201 Charles Street, Providence, RI 02904-2294. Periodicals postage paid at Providence, RI. Postmaster: Send address changes to Memoirs, American Mathematical Society, P. O. Box 6248, Providence, RI 02940-6248.

This publication is indexed in *Science Citation Index*®, *SciSearch*®, *Research Alert*®, *CompuMath Citation Index*®, *Current Contents*®*/Physical, Chemical & Earth Sciences.*
Printed in the United States of America.

∞ The paper used in this book is acid-free and falls within the guidelines established to ensure permanence and durability.
Visit the AMS home page at URL: `http://www.ams.org/`

10 9 8 7 6 5 4 3 2 1 07 06 05 04 03 02

Contents

Abstract

We consider the concept of triangulation of an oriented matroid. We provide a definition which generalizes the previous ones by Billera–Munson and by Anderson and which specializes to the usual notion of triangulation (or simplicial fan) in the realizable case.

Then we study the relation existing between triangulations of an oriented matroid $\mathcal{M}$ and extensions of its dual $\mathcal{M}^*$, via the so-called *lifting triangulations*. We show that this duality behaves particularly well in the class of *Lawrence matroid polytopes*. In particular, that the *extension space conjecture* for realizable oriented matroids is equivalent to the restriction to Lawrence polytopes of the Generalized Baues problem for subdivisions of polytopes.

We finish by showing examples and a characterization of lifting triangulations.

Received by the editor July 1997, and with revisions December 1999 .

2000 *Mathematics Subject Classification.* Primary 52C40; Secondary 52B11, 52B35.

Key words and phrases. Oriented matroid, triangulation, polytope, combinatorial convexity.

Partially supported by grant PB97-0358 of the Spanish Dirección General de Investigación Científica y Técnica.

Most of this work was done during 1996, while the author stayed at the Mathematical Institute, University of Oxford, as a Postdoctoral Fellow under contract ERBCHBGCT940647 of the European Union.

I am grateful to Bernd Sturmfels, who proposed me the study of oriented matroid triangulations as a natural continuation of the results in [**14**]. To Jörg Rambau, who clarified my initial ideas on euclideanness and circuit intersection. And to Jesús de Loera, who carefully read several preliminary versions of the manuscript and helped with suggestions, comments and encouragement. An anonymous referee thoroughly read and commented the first version submitted to the editors.

Introduction

Matroids (see [**27**]) and oriented matroids (see [**11**]) are axiomatic abstract models for combinatorial geometry over general fields and over ordered fields, respectively. Oriented matroids have some extra structure as compared to matroids, one of whose features is the existence of a notion of *convexity* (see Chapter 9 of [**11**]). This makes it natural to consider triangulations of oriented matroids as an analogue of triangulations of usual polytopes or point configurations. This concept is the object of this paper.

Triangulations of oriented matroids generalize the following situations, where the oriented matroids involved are realized by geometric objects and where the triangulations of the geometric objects considered are known to depend only on the underlying oriented matroid:

- If $\mathcal{M}$ is the oriented matroid of affine dependences between the vertices of a polytope P, the triangulations of $\mathcal{M}$ coincide with the triangulations of the polytope P, meaning by this the geometric simplicial complexes which cover P and use only the vertices of P as vertices. There is a recent survey by Lee [**25**] on this topic.
- If $\mathcal{M}$ is the oriented matroid of affine dependences between the points in a finite point set $\mathcal{A}$ in $\mathbb{R}^d$, the triangulations of $\mathcal{M}$ coincide with the triangulations of $\mathcal{A}$, meaning by this the geometric simplicial complexes which cover the convex hull of $\mathcal{A}$ and which use (perhaps not all) the points of $\mathcal{A}$ as vertices. This is a generalization of the previous case which has been often considered in recent literature (see [**6, 13, 14, 15, 16, 28, 30, 36**] and Chapter 7 of [**19**]).
- If $\mathcal{M}$ is the oriented matroid of linear dependences between a finite set of vectors $\mathcal{V}$ in $\mathbb{R}^d$, the triangulations of $\mathcal{M}$ coincide with the triangulations of $\mathcal{V}$, meaning by this the simplicial fans covering the positive span of $\mathcal{V}$ and whose rank-1 cones are generated by (perhaps not all) the vectors of $\mathcal{A}$. See [**19**, Definition 4.1] or [**7**]. This is a further generalization of the previous case: if $\mathcal{A}$ is a point configuration in $\mathbb{R}^d$, then triangulations of $\mathcal{A}$ coincide with the simplicial fans of the vector configuration obtained by embedding $\mathbb{R}^d$ as an affine hyperplane in the vector space $\mathbb{R}^{d+1}$.

Triangulations of oriented matroids were first defined by Billera and Munson in [**9**], for the special case of *polytopal* oriented matroids. An account of them for the more general case of *acyclic* oriented matroids appears in Section 9.6 of [**11**].

Deciding which is the best, or natural, definition of triangulation of a general oriented matroid is not trivial. Apart of the problem of translating geometric conditions into oriented matroid language, different possible characterizations of triangulations in the realized case may translate to non-equivalent definitions in oriented matroid terms. For example, Billera and Munson [**9**] consider two possible definitions, one apparently stronger than the other, and work with the weaker one for practical reasons although they admit the stronger one to be a perhaps more direct translation of the usual definition of triangulation of a polytope (see Remark 2.5(ii)). One of our first results in this paper says that the two definitions were in fact equivalent. Actually, in Theorem 2.4 we give seven different characterizations of oriented matroid triangulations, which include (more precisely, generalize, since we do not assume our oriented matroids to be neither acyclic nor totally cyclic) the two definitions by Billera and Munson and the recursive one given later by Anderson in [**1**]. Anderson is primarily interested in the totally cyclic case, although she proves that the polytopal version of her definition is equivalent to the weak one of Billera and Munson.

After the definition problem is solved, the goals of this paper are two-fold. On the one hand we generalize to triangulations of perhaps non-realizable oriented matroids results which are known for triangulations of point or vector configurations. For example, Sections 3.1 and 3.2 generalize most of the results in Sections 2 and 5 of [**14**]. On the other hand, we use the insight provided by oriented matroids to obtain results which are new even in the realized case. For example, the main result of Chapter 4 is that the extension space of an oriented matroid is isomorphic to the poset of proper subdivisions of a Lawrence polytope. This shows the equivalence of two open cases of the generalized Baues conjecture (see below).

⋆

One motivation for studying triangulations of non-realizable oriented matroids comes from the theory of *combinatorial differential manifolds* introduced by MacPherson [**26**], see also [**2**]. These manifolds provide a promising interplay between differential and combinatorial geometry, one of whose first outcomes has been the combinatorial formula for the Pontrjagin classes of triangulated differential manifolds obtained by Gel'fand and MacPherson in [**20**].

The connection between differential combinatorial manifolds and triangulations of oriented matroids was exhibited by Anderson [**1**]. Roughly speaking, combinatorial differential manifolds are simplicial topological pseudo-manifolds with a combinatorial analogue to a differential structure defined by means of oriented matroids. Anderson has proved that the link of every cell of a combinatorial differential manifold is an oriented matroid triangulation. In particular, the topology type problem for triangulations of oriented matroids (see the open problems at the end of this introduction, or Section 2.4) is equivalent to deciding whether all differential combinatorial manifolds are really manifolds.

Concerning triangulations of realized oriented matroids, the problems we have in mind have to do with the notion of *bistellar flip* – a sort of "elementary move" between triangulations – and the order poset of subdivisions of a polytope or point configuration. Both objects have received attention in recent literature, partially as a result of the theory of *secondary polytopes* developed by Gel'fand, Kapranov and Zelevinsky (see Chapter 7 of [**19**], in particular pages 231–233 for the original definition of bistellar flip). The following two questions are natural: has the order complex of polytopal subdivisions of a point configuration with n vertices in dimension d the homotopy type of a $(n-d-2)$-sphere? Is any pair of triangulations connected by a sequence of bistellar flips?

The first question is still open, and is a special case of the *generalized Baues conjecture* posed by Billera et al. [**8**] (the conjecture in its full generality was disproved in [**31**]; see [**32**] for a survey on this problem). The second is a weak version of it, for which a counter-example has been recently exhibited by the author [**36**]. Previous to that, only very few cases had been answered, always in the affirmative: The case of "few points" ($n \leq d+3$) is answered by the results of Lee [**24**] and the theory of secondary polytopes [**19**, Chapter 7] and [**6**]. Azaola [**3**] has recently proved the case $n = d+4$. In the case of "low dimension" ($d \leq 2$) the connectivity question is known since some time ago [**23**] while the homotopy question has been solved by Edelman and Reiner [**16**]. A particular case interesting because of the amount of extra combinatorial structure available is that of *cyclic polytopes*, solved in [**28**] and [**30**].

The most interesting new result for the realizable case obtained in this paper is that we prove in Section 4.3 (see Corollary 4.16) the equivalence of the following two open cases of the generalized Baues conjecture. After this paper was finished a different, geometric, proof of this same equivalence has appeared in [**21**]:

- The *extension space conjecture* of oriented matroid theory (see [**11**, pages 295–296] or [**39**]), stating that the poset of non-trivial single-element extensions of any rank r realizable oriented matroid has the homotopy type of an $(r-1)$-sphere. Via the Bohne-Dress theorem on zonotopal tilings (see [**11**, Theorem 2.2.13] or [**40**, Theorem 7.32]) this was previously known to be equivalent to the conjecture that the poset of proper zonotopal subdivisions of any zonotope with n generators and dimension $n-r$ has this same homotopy type. I.e., to the generalized Baues conjecture for the case of projections from a cube.
- The conjecture that the poset of all proper subdivisions of any (realized) Lawrence polytope of dimension d with n vertices is homotopy equivalent to an $(n-d-2)$-sphere. I.e., to the generalized Baues conjecture in the particular case of the projection from a simplex to a Lawrence polytope.

⋆

Summarizing, the main results of the paper are:

(a) We settle the problem of defining triangulations of oriented matroids by providing a definition which suits any oriented matroid and giving several different characterizations of it (Theorem 2.4).

(b) We generalize to non-realizable oriented matroids the results in [**14**] concerning the duality between triangulations of a configuration and *chambers* (related to extensions, in the oriented matroid setting) of the dual configuration (Sections 3.1 and 3.2). Under this duality, bistellar flips between triangulations are related to mutations between extensions (Theorem 3.14).

(c) We prove that in the case of Lawrence polytopes the duality behaves specially well, because the correspondence between triangulations of a Lawrence polytope and extensions of its dual oriented matroid is bijective and makes bistellar flips correspond exactly to mutations (Theorem 4.14).

(d) In particular, this implies that the *extension space conjecture* of oriented matroid theory is equivalent to the Generalized Baues problem for realized Lawrence polytopes (Corollary 4.16). Also, that there are (non-realizable) Lawrence polytopes of corank four whose triangulations are not connected by bistellar flips (Corollary 4.15).
(e) We introduce a reoriented version of the Lawrence construction (Section 4.4, Theorem 4.18), which shows that the "polytopal case" cannot be considered simpler than the "acyclic case" when dealing with triangulations, unless the dimension is fixed.
(f) Not all the triangulations of an oriented matroid correspond to extensions of the dual. The ones which do correspond are called *lifting* (see [**11**, Section 9.6], or our Section 4.1). We give necessary conditions (Proposition 5.3) and two characterizations (Theorem 5.10) for an oriented matroid triangulation to be a lifting triangulation. Actually, these conditions are proved in the more general case of lifting subdivisions.
(g) We construct non-lifting triangulations of the 4-cube and of a unimodular polytope (Sections 5.2.3 and 5.2.2). We also show bad behavior of triangulations of non-Euclidean oriented matroids, with an example in the Edmonds-Fukuda-Mandel oriented matroid `EFM`(8) (Section 5.2.1).

The following is a more detailed description of some of these points, and of the structure of the paper:

The technical tools from oriented matroid theory that we will need concern mainly single-element extensions and convexity. That is, the first part of Chapter 7 and Chapter 9 in [**11**]. In Chapter 1 we recall these concepts and prove some preliminary results which will be frequently used. Readers familiar with oriented matroid theory can skip this section and come to it only for reference. Other readers may find this section useful for understanding the convex geometry of oriented matroids.

In Sections 2.1 and 2.2 we give our definition of oriented matroid triangulation, based in the "weak" one by Billera and Munson, and provide the above mentioned equivalent characterizations of it (Theorem 2.4). In Section 2.3 we prove properties of triangulations which will be needed afterwards. Section 2.4 states what is known about the topological type of triangulations of oriented matroids, based mostly on [**1**] and on personal communications by J. Rambau.

In Section 3.1 we introduce the duality existing between *lifting* triangulations of an oriented matroid and extensions in general position of its dual. Lifting triangulations were introduced in [**11**, Section 9.6] (a particular case was mentioned in [**9**]). We give two equivalent definitions of them, dual to one another (definitions 3.4 and 4.1).

This duality is a generalization of the duality between *regular triangulations* of a point configuration $\mathcal{A}$ and *chambers* of its Gale transform $\mathcal{A}^*$, exhibited by Billera et al. in [**6**]. De Loera et al. [**14**] have already given a generalized version of this duality, still in the realizable case, with the introduction of *virtual chambers*. Section 3.2 is the translation into the oriented matroid setting of Sections 2 and 5 of [**14**], and some of the proofs required few changes. In particular, we show that for any triangulations T of $\mathcal{M}$ and T' of the dual $\mathcal{M}'$ there are unique maximal simplices of T and T' which are complements (Theorem 3.8). In Section 3.3 we show the exact relation between the natural notions of elementary change on triangulations (the notion of *bistellar flip*) and on extensions (the notion of *mutation*) under the duality. Namely, that whenever two extensions differ by a mutation the corresponding lifting triangulations either coincide or differ by a bistellar flip (Theorem 3.14).

Subdivisions of oriented matroids are a generalization of triangulations introduced in [**11**, Section 9.6]. We devote Section 4.1 to study them in some detail. Going further on the duality mentioned above, it is easy to establish a surjective order-preserving map from the poset of extensions of an oriented matroid ordered by weak maps and the poset of *lifting subdivisions* of the dual oriented matroid ordered by refinement (see Exercises 9.30 and 9.31 in [**11**]). Section 4.2 is devoted to the specially nice case of *Lawrence polytopes*, leading to the isomorphism between the two posets mentioned. Section 4.3 describes the consequences of this in the context of the Baues problem.

In Section 4.4 we introduce a reoriented version of the Lawrence construction, which translates results on triangulations of polytopes to non-polytopal point configurations and vice-versa.

Since lifting triangulations have played an important role in the results so far, we devote to them the last chapter. We show some good properties of them and their relation with regular triangulations of point configurations in Section 5.1. We construct interesting examples of non-lifting triangulations in Section 5.2. In Section 5.3 we prove the following surprising fact: although the definition of liftingness for a triangulation or subdivision relies strongly in the notion of oriented matroid, there are two "oriented-matroid-free" characterizations of liftingness for triangulations of a realized oriented matroid (Theorem 5.10): lifting subdivisions of $\mathcal{M}$ are the links of subdivisions of the Lawrence polytope over $\mathcal{M}$ and they are also the ones which are compatible with subdivisions of every restriction of $\mathcal{M}$, in a sense specified in Definition 5.9. This seems to imply that liftingness is a natural concept even outside oriented matroid theory. The second characterization has been used in [**30**] to prove that cyclic polytopes only have lifting triangulations and in [**21**] to provide a different proof of our Theorem 4.14 and Corollary 4.16.

⋆

We finish this introduction pointing out some open problems:

- We have proved the equivalence of the extension space conjecture to the conjecture on the order complex of subdivisions of a realizable Lawrence polytope. But it remains to find out whether they are both true or both false. See the details in Section 4.3.
- What is the topological type of an oriented matroid triangulation, considered as a simplicial complex? The expected answer is that a triangulation of an oriented matroid $\mathcal{M}$ of rank r is homeomorphic to an $(r-1)$-ball if $\mathcal{M}$ is totally cyclic and to an $(r-1)$-sphere otherwise. No proof of this exists in the general case, although it holds trivially for realizable oriented matroids and, by [**11**, Proposition 9.1.1], for lifting triangulations of arbitrary oriented matroids. Anderson [**1**] has provided a proof for *Euclidean* oriented matroids (see Theorem 2.16). We recall that euclideanness is a property satisfied by realizable oriented matroids and which can be rephrased vaguely as "every line and hyperplane meet in a point" (see Section 1.3, or [**11**, Sections 7.5 and 10.5]).

- There are other properties of triangulations which are known to hold under some euclideanness assumption, but open in general:
  - Can two simplices of a triangulation contain respectively the positive and negative parts of a circuit? That the answer is "no" appeared as Conjecture 6.01 in Laura Anderson's Ph. D. thesis, on which [**1**] is based. See Remark 2.5(v).
  - In Definition 2.7 we construct a graph which tries to mimic the sequence of simplices intersected by a segment in a triangulation of a point configuration. This graph was first used by Anderson [**1**]. Can the graph have cycles? See the comments after Lemma 2.8 and Section 2.4. Proposition 2.17 (by Rambau) relates this question to the previous one.
  - We say that two subsets A and B of the ground set E of an oriented matroid are *separated* if there is a covector (C^+, C^-) with $A \cap C^+ = B \cap C^- = \emptyset$ and such that $A \cap C^0$ and $B \cap C^0$ are separated in the (lower rank) oriented matroid $\mathcal{M}(C^0)$ (here, C^0 denotes $E \setminus (C^+ \cup C^-)$). Can every separated pair of simplices be completed to a triangulation? In the presence of some euclideanness condition, every pair of separated simplices belongs to some *lifting* triangulation. In Proposition 5.6 we show two separated simplices which do not belong to any lifting triangulation, but they belong to a non-lifting one.
- All the non-lifting subdivisions which appear in this paper have the property that their restriction to some minor cannot be extended to a subdivision of that minor (which proves that they are non-lifting, by Proposition 5.3). Is this a general property of non-lifting subdivisions? If yes, then every subdivision of a rank 3 oriented matroid will be lifting.
- In [**35**] the author proves that every non-regular subdivision can be refined to a non-regular triangulation. Can every non-lifting subdivision be refined to a non-lifting triangulation? This is not obvious; for example, the proof of liftingness for all triangulations of cyclic polytopes given in [**30**] does not extend to subdivisions. Is every subdivision of a cyclic polytope lifting?

CHAPTER 1

Preliminaries on Oriented Matroids

Throughout the paper we will assume familiarity with the basics of oriented matroid theory. In this section we sum up the main specific concepts and properties that we will need, related mainly to *convexity* and *extensions*. We will follow the book by Björner et al. [**11**] for notation and reference, unless otherwise indicated.

Since we will be very seldom concerned with (non-oriented) matroids, we will use the terms *circuits*, *cocircuits*, *vectors* and *covectors* always referring to *signed* ones. We will indistinctly consider them signed subsets $C = (C^+, C^-)$ of E or functions $C : E \longrightarrow \{-1, 0, +1\}$, where E is the ground set of the oriented matroid. Using the second point of view we can say that a circuit "is positive" or that it "vanishes" at some elements of E, and will write $C(p) = +1$ with the same meaning as $p \in C^+$, for $p \in E$. As usual, $\underline{C}$ denotes the support $C^+ \cup C^-$ of C.

1.1. Convexity.

Let $\mathcal{M}$ be an oriented matroid of rank r on a set E. In order to stress the geometrical meaning of oriented matroid concepts we will call *simplices* of $\mathcal{M}$ the independent subsets of E. A k-simplex is a simplex with k-elements. Thus, r-simplices are the same thing as bases. If $\mathcal{M}$ is a realizable oriented matroid and $\mathcal{V} \subseteq \mathbb{R}^r$ is a vector realization of $\mathcal{M}$ then the geometric counterpart of the k-simplices of $\mathcal{M}$ are simplicial cones of dimension k positively spanned by independent subsets of $\mathcal{M}$. If $\mathcal{M}$ is acyclic and realized by a point configuration $\mathcal{A} \subseteq \mathbb{R}^{r-1}$ then the k-simplices of $\mathcal{M}$ correspond to simplices of $\mathcal{A}$ of dimension $k-1$, with vertex set contained in $\mathcal{A}$.

Following [**11**, Chapter 9], we call *facets* of $\mathcal{M}$ the complements of supports of non-negative cocircuits of $\mathcal{M}$ and *faces* the complements of supports of non-negative covectors. Facets are the maximal *proper* faces (faces different from E itself). In contrast with [**11**], we do not assume $\mathcal{M}$ to be acyclic. The faces of an oriented matroid form a lattice called the *Las Vergnas face lattice*. The unique maximal face is E and the unique minimal face is the family F_0 of elements which lie in positive circuits. It is the empty set if $\mathcal{M}$ is acyclic. If $\mathcal{M}$ is totally cyclic then $F_0 = E$ and $\mathcal{M}$ has no proper faces.

For any $A \subseteq E$ we denote by $\mathcal{M}(A)$ the restriction of $\mathcal{M}$ to A. We call faces (resp. facets) of A the faces (resp. facets) of $\mathcal{M}(A)$. In particular, every subset of a k-simplex is a face of it, and it is a facet if and only if it has $k-1$ elements. Of course, all the faces of a simplex are simplices.

The *convex hull* of a subset $A \subseteq E$ is the union of A and those elements p of $E \setminus A$ for which there is a signed circuit C of $\mathcal{M}$ with $C^+ = \{p\}$ and $C^- \subseteq A$. We denote this set by $\mathrm{conv}_{\mathcal{M}}(A)$. The *relative interior* of A is the set obtained removing the convex hulls of facets of A from the convex hull of A. We denote it $\text{relint}_{\mathcal{M}}(A)$.

LEMMA 1.1. *Let $\mathcal{M}$ be an oriented matroid of rank r on a set E. Let $p \in E$ and $A \subseteq E$. Then:*

(i) *$p \in conv_{\mathcal{M}}(A)$ if and only if $p \in conv_{\mathcal{M}(A \cup p)}(A)$, where $\mathcal{M}(A \cup p)$ is the restriction of $\mathcal{M}$ to $A \cup p$.*
(ii) *if $rank_{\mathcal{M}}(A) = k$, then $p \in conv_{\mathcal{M}}(A)$ if and only if there is a k-simplex $\tau \subseteq A$ with $p \in conv_{\mathcal{M}}(\tau)$.*
(iii) *$p \in conv_{\mathcal{M}}(A)$ if and only if every cocircuit of $\mathcal{M}$ which is nonnegative on A is nonnegative at p.*
(iv) *if $p \in conv_{\mathcal{M}}(A)$ and $A \subseteq conv_{\mathcal{M}}(B)$ $(B \subseteq E)$ then $p \in conv_{\mathcal{M}}(B)$.*
(v) *if A is an r-simplex, then $p \in conv_{\mathcal{M}}(A)$ if and only if for every $a \in A$ the unique cocircuit of $\mathcal{M}$ vanishing on $A \backslash a$ and positive on a is non-negative at p.*

PROOF. Part (i) follows from the fact that circuits of $\mathcal{M}(A \cup p)$ correspond exactly with circuits of $\mathcal{M}$ with support contained in $A \cup p$. Part (ii), considered on the oriented matroid $\mathcal{M}(A \cup p)$, is "Carathéodory's Theorem" [**11**, Theorem 9.2.1(1)]. Part (iii) is "Weyl's Theorem" [**11**, Theorem 9.2.1(2)]. Part (iv) follows from (iii).

The "only-if" part in (v) is a consequence of (iii). For the "if" part consider a circuit C with support contained in the spanning but not independent subset $A \cup \{p\}$. Since A is independent, p is in the support of C and without loss of generality we assume $C(p) = +1$. If C was positive at some element $a \in A$, then the orthogonality of C and the cocircuit vanishing on $A \setminus a$ would be violated in the restricted oriented matroid $\mathcal{M}(A \cup \{p\})$. □

LEMMA 1.2. *Let $\mathcal{M}$ be an oriented matroid on a set E. Let $a \in E$ and $A, B \subseteq E$. Then:*

(i) *$p \in relint_{\mathcal{M}}(A)$ if and only if $p \in conv_{\mathcal{M}}(A)$ and for every covector $C = (C^+, C^-)$ vanishing on p, either C vanishes on A or has both negative and positive elements in A.*
(ii) *if $p \in relint_{\mathcal{M}}(A)$ and $A \subseteq conv_{\mathcal{M}}(B)$, but A is not contained in the convex hull of any facet of B, then $p \in relint_{\mathcal{M}}(B)$.*
(iii) *if A is an independent set, then $p \in relint_{\mathcal{M}}(A)$ if and only if $(\{p\}, A)$ is a circuit of $\mathcal{M}$.*

PROOF. If $p \in relint_{\mathcal{M}}(A)$, then any cocircuit which is nonnegative on A either vanishes on A or does not vanish at p; otherwise p will be in the convex hull of a facet of $\mathcal{A}$, or not in the convex hull of A. Reciprocally, if $p \in conv_{\mathcal{M}}(A)$, but $p \notin relint_{\mathcal{M}}(A)$, then there is a cocircuit which is nonnegative on A, vanishes at p and does not vanish on A. This proves (i).

For (ii), consider a cocircuit C vanishing at p but not on all of B. If C vanishes on A, as A is not in a facet of B, C takes both signs on B. If C does not vanish on A, it takes both signs on A and, hence, also on B (because $A \subseteq conv_{\mathcal{M}}(B)$, and using part (iii) of Lemma 1.1). We conclude that $p \in relint_{\mathcal{M}}(B)$, by the characterization in part (i).

In (iii), A being independent implies that there is at most one circuit with support contained in $A \cup p$. By definition of convex hull, $p \in conv_{\mathcal{M}}(A)$ if and only if the circuit is of the form $(\{p\}, B)$ for some $B \subseteq A$. If this is the case we have two possibilities: if $B \neq A$, then p is in the convex hull of a proper face of A, and thus not in $relint_{\mathcal{M}}(A)$. If $B = A$, then the orthogonality between circuits and covectors implies, with part (i), that $p \in relint_{\mathcal{M}}(A)$. □

1.2. Extensions. Lexicographic extensions.

Let $\mathcal{M}$ and $\mathcal{M}'$ be two oriented matroids on sets E and E'. If $E \subseteq E'$, and every circuit of $\mathcal{M}$ is a circuit in $\mathcal{M}'$ then $\mathcal{M}'$ is an *extension* of $\mathcal{M}$. Equivalently, $\mathcal{M}'$ is an extension of $\mathcal{M}$ if $\mathcal{M}$ is a restriction of $\mathcal{M}'$ by deleting some elements. We will only consider extensions which do not increase the rank, i.e., for which $rank(\mathcal{M}) = rank(\mathcal{M}')$. If $E' \setminus E = \{p\}$ has one element we say that $\mathcal{M}'$ is a *one-element* extension, and use the notation $\mathcal{M} \cup p$ for $\mathcal{M}'$. This will be usually our case.

Let $\mathcal{M} \cup p$ be a one-element extension of $\mathcal{M}$. For every cocircuit $C = (C^+, C^-)$ of $\mathcal{M}$, exactly one of $(C^+ \cup \{p\}, C^-)$, $(C^+, C^- \cup \{p\})$ and (C^+, C^-) is a cocircuit of $\mathcal{M}$. In other words, there is a unique way to extend each cocircuit of $\mathcal{M}$ into a cocircuit of $\mathcal{M} \cup p$. This means that there is no ambiguity in considering C as a cocircuit in $\mathcal{M} \cup p$, and we can write $C(p) = +1, -1$ and 0, respectively. The function assigning to each cocircuit of $\mathcal{M}$ its value $C(p) \in \{-1, 0, +1\}$ on the new element p is called the *signature* of the extension $\mathcal{M} \cup p$.

Not every map from the set of cocircuits of $\mathcal{M}$ to $\{-1, 0, +1\}$ is the signature function of an extension. Also, not every cocircuit of an extension $\mathcal{M} \cup p$ is the extension of a cocircuit of $\mathcal{M}$. However, it is true that a valid signature function on the cocircuits of $\mathcal{M}$ uniquely determines the extension. More information on this can be found in [**11**, Section 7.1]. In particular, the way to obtain all the cocircuits of $\mathcal{M} \cup p$ from the cocircuits of $\mathcal{M}$ and the signature function of $\mathcal{M} \cup p$ is in Proposition 7.1.4. The conditions that a signature function has to satisfy to be valid are in Theorem 7.1.8, which we reproduce below. Both results come from the paper [**22**] by Las Vergnas.

LEMMA 1.3 (Las Vergnas). *Let $\mathcal{M}$ be an oriented matroid on a ground set E, and $\mathcal{C}^*$ its set of cocircuits. Let $\sigma : \mathcal{C}^* \to \{+, -, 0\}$ be a cocircuit signature satisfying $\sigma(-C) = -\sigma(C)$ for every $C \in \mathcal{C}^*$. Then, the following conditions are equivalent:*

(a) *σ is the cocircuit signature function of a single-element extension of $\mathcal{M}$.*
(b) *For every subset $A \subseteq E$ of corank 2, σ restricted to the cocircuits not intersecting A is the cocircuit signature function of an extension of $\mathcal{M}/A$. I.e., σ defines a single-element extension on every corank 2 contraction.*
(c) *$\mathcal{M}$ has no minor of rank 2 on three elements on which σ induces one of the three forbidden subconfigurations displayed in Figure 1.1. (The figure should be read as follows: the three lines represent the complement of cocircuits, i.e. flats of rank 1 of $\mathcal{M}$. A plus or minus sign on one side of a flat mean that p lies in this or the other part of the cocircuit, respectively. A zero means that p lies on the flat).* □

DEFINITION 1.4. Let $\mathcal{M} \cup p$ be a one-element extension of an oriented matroid $\mathcal{M}$ of rank r on a set E. We say that the extension is *interior* if $p \in conv_{\mathcal{M} \cup p}(E)$. We say that the extension is in *general position* if $C(p) \neq 0$, for every cocircuit C of $\mathcal{M}$; equivalently, if the support of every circuit of $\mathcal{M} \cup p$ containing p is a spanning subset.

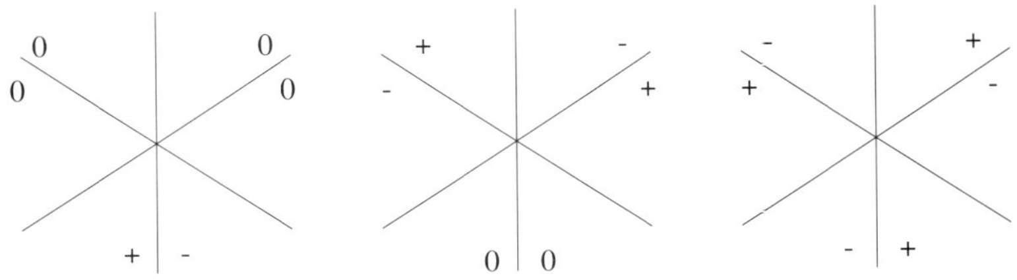

FIGURE 1.1. Forbidden subconfigurations for cocircuit signatures.

DEFINITION 1.5. Let $\mathcal{M} \cup p$ and $\mathcal{M} \cup p'$ be two one-element extensions of an oriented matroid $\mathcal{M}$. We say that p' is a perturbation of p if there is a weak map from $\mathcal{M} \cup p$ to $\mathcal{M} \cup p'$. Equivalently, if for every cocircuit C of $\mathcal{M}$

$$C(p) \neq 0 \qquad \Longrightarrow \qquad C(p) = C(p')$$

We will be particularly interested in the so-called *lexicographic extensions*, which were also introduced by Las Vergnas. We take as a definition the following characterization of them which appears in [**11**, Proposition 7.2.4].

DEFINITION 1.6. Let $\mathcal{M}$ be an oriented matroid on a set E. Let $\{a_1, \dots, a_k\}$ be a subset of E and choose a sign $\epsilon_i \in \{+, -\}$ for each $i = 1, \dots, k$. The lexicographic extension $\mathcal{M} \cup p$ of $\mathcal{M}$ by the element $p := [a_1^{\epsilon_1}, \dots, a_k^{\epsilon_k}]$ is defined as the one whose cocircuit signature is given by:

$$C(p) = \begin{cases} \epsilon_i C(a_i) & \text{if } i \text{ is minimal with } C(a_i) \neq 0 \\ 0 & \text{if } C(a_i) = 0, \quad \forall i = 1, \dots, k \end{cases}$$

If $\epsilon_i = +$ for all i then we call $\mathcal{M} \cup p$ a *positive lexicographic extension.*

In particular, with $p := [a^+]$ we obtain the extension by an element p parallel to a, and with $p := [a^-]$ the extension by an element antiparallel to a. In the definition, there is no loss of generality if we assume $a_1, \dots, a_k$ to be independent. In fact, if l is the first index for which $a_1, \dots, a_l$ is dependent, then the element a_l can be removed from the definition without affecting the extension obtained.

For another example, let $\mathcal{M} \cup p$ be any one-element extension of $\mathcal{M}$ and consider the lexicographic extension of $\mathcal{M} \cup p$ by an element $p' := [p^+, a_1^{\epsilon_1}, \dots, a_k^{\epsilon_k}]$. Then, the extension $\mathcal{M} \cup p' := (\mathcal{M} \cup p) \cup p' \setminus p$ is a perturbation of $\mathcal{M} \cup p$ (In [**11**, p. 292] this extension is called the *composition* of p_1 and the lexicographic extension $[a_1^{\epsilon_1}, \dots, a_k^{\epsilon_k}]$). In case that p itself is a lexicographic extension $p := [b_1^{\delta_1}, \dots, b_l^{\delta_l}]$, then p' is the lexicographic extension given by $p' := [b_1^{\delta_1}, \dots, b_l^{\delta_l}, a_1^{\epsilon_1}, \dots, a_k^{\epsilon_k}]$.

LEMMA 1.7. *Let $\mathcal{M}$ be an oriented matroid of rank r. Let $\mathcal{M} \cup p$ be a lexicographic extension of $\mathcal{M}$ by the element $p := [a_1^{\epsilon_1}, \dots, a_k^{\epsilon_k}]$. Then,*

(i) *the extension is in general position if and only if $rank(\{a_1, \dots, a_k\}) = r$.*
(ii) *if $\epsilon_i = +$ for all i, then $p \in relint_{\mathcal{M} \cup p}(\{a_1, \dots, a_k\})$. In particular, it is an interior extension.*

PROOF. Both parts are special cases of [**11**, Lemma 7.2.6], originally proved by Todd in 1984. □

LEMMA 1.8. *Let $\mathcal{M}$ be an oriented matroid of rank r on a set E. Let $A \subseteq E$ and let $\mathcal{M} \cup p$ and $\mathcal{M} \cup p'$ be two extensions of $\mathcal{M}$, with p' being a perturbation of p. Then:*

(i) *if p'' is a perturbation of p', then p'' is a perturbation of p as well.*
(ii) *if $p' \in conv_{\mathcal{M} \cup p'}(A)$, then $p \in conv_{\mathcal{M} \cup p}(A)$.*
(iii) *if A has rank r and $p \in relint_{\mathcal{M} \cup p}(A)$, then $p' \in relint_{\mathcal{M} \cup p'}(A)$.*
(iv) *if $p \in conv_{\mathcal{M} \cup p}(A)$ and $p' := [p^+, a_1^+, \ldots, a_k^+]$, then $p' \in conv_{\mathcal{M} \cup p'}(A \cup \{a_1, \ldots, a_k\})$.*

PROOF. Part (i) is obvious from the definition. Part (ii) follows from the characterization of the convex hull by cocircuits (part (iii) of Lemma 1.1). Part (iii) follows from the characterization of the relative interior by cocircuits (part (i) of Lemma 1.2) and the fact that every cocircuit of $\mathcal{M}$ which vanishes on p' also vanishes at p. Part (iv) follows from the fact that, in the oriented matroid $\mathcal{M} \cup \{p, p'\}$, p' is in the convex hull of $\{p, a_1, \ldots . a_k\}$ (part (ii) of Lemma 1.7) and p is in the convex hull of A. □

An important difficulty concerning extensions is that given an oriented matroid $\mathcal{M}$ on a set E, a subset $A \subseteq E$, and an extension $\mathcal{M}(A) \cup p$ of a restriction of $\mathcal{M}$, there might not exist an extension $\mathcal{M} \cup p$ of $\mathcal{M}$ which extends the given one. By this we mean that $(\mathcal{M} \cup p)(\mathcal{A} \cup \{p\}) = \mathcal{M}(A) \cup p$.

For example, the oriented matroid $\mathcal{M}$ realized as a point configuration by the six vertices of a convex hexagon has two extensions which are incompatible: the extension $\mathcal{M} \cup p_1$ by an element lying in the intersection of the three main diagonals and an extension $\mathcal{M} \cup p_2$ by an element lying in the intersection of two of them, but not on the third one. This means that the extension p_2 of $\mathcal{M}$ can not be extended to $\mathcal{M} \cup p_1$. Our next result implies that this bad behaviour does not occur if one of the extensions is lexicographic. After that we prove that the corresponding bad behaviour for contractions never arises.

LEMMA 1.9. *Let $\mathcal{M}$ be an oriented matroid on a set E and let $A \subseteq E$. Let $\mathcal{M}(A) \cup p$ be a lexicographic extension of the restriction $\mathcal{M}(A)$. Let $\mathcal{M} \cup p$ denote the lexicographic extension of $\mathcal{M}$ with the same lexicographic expression. Then:*

(i) *$(\mathcal{M} \cup p)(A \cup \{p\}) = \mathcal{M}(A) \cup p$.*
(ii) *if p is interior in $\mathcal{M}(A)$, then p is interior in $\mathcal{M}$.*
(iii) *if p is in general position and A spans $\mathcal{M}$, then p' is in general position.*

PROOF. Every cocircuit of $\mathcal{M}(A)$ extends to a cocircuit of $\mathcal{M}$. The fact that p has the same lexicographic expression in $\mathcal{M}$ and $\mathcal{M}(A)$ implies that its cocircuit signature on those cocircuits of $\mathcal{M}$ coincides with the one it has in $\mathcal{M}(A)$. Hence $(\mathcal{M} \cup p)(A \cup \{p\}) = \mathcal{M}(A) \cup p$.

Part (ii) is a direct consequence of parts (i) and (iv) of Lemma 1.1. Part (iii) follows from part (i) of Lemma 1.7. □

LEMMA 1.10. *Let $\mathcal{M}$ be an oriented matroid on a set E and let $a \in E$. Let $(\mathcal{M}/a) \cup p$ be an extension of the contraction $\mathcal{M}/a$. Every cocircuit of $\mathcal{M}$ which vanishes on a is a cocircuit of $\mathcal{M}/a$; thus, the following cocircuit signature on $\mathcal{M}$ is well defined: $C(p') = C(a)$ if $C(a) \neq 0$ and $C(p') = C(p)$ otherwise. Then,*

(i) *The cocircuit signature defines an extension $\mathcal{M} \cup p'$ of $\mathcal{M}$ which satisfies $(\mathcal{M} \cup p')/a = (\mathcal{M}/a) \cup p$.*

(ii) *if p is interior in $\mathcal{M}/a$, then p' is interior in $\mathcal{M}$.*
(iii) *if p is in general position, then p' is in general position.*
(iv) *if p is lexicographic, defined by an expression $[a_1^{\epsilon_1}, \dots a_k^{\epsilon_k}]$, then p' is the lexicographic extension defined by the expression $[a^+, a_1^{\epsilon_1}, \dots a_k^{\epsilon_k}]$.*
(v) *p' is the only perturbation of the extension of $\mathcal{M}$ by an element which satisfies $(\mathcal{M} \cup p')/a = (\mathcal{M}/a) \cup p$.*
(vi) *For every subset $A \subseteq E \setminus \{a\}$, we have*

$$p \in conv_{(\mathcal{M}/a)\cup p}(A) \qquad \Longleftrightarrow \qquad p' \in conv_{\mathcal{M}\cup p'}(A \cup \{a\}).$$

PROOF. The lexicographic case stated in part (iv) is trivial. For the general case, let $\mathcal{M}/A$ be a rank-2 contraction of $\mathcal{M}$. $\mathcal{M}/(A \cup \{a\})$ has rank at most 2 and its extension $((\mathcal{M}/a) \cup p)(A \cup \{a\})$ is a lexicographic extension. (every extension in rank at most 2 is lexicographic) By the lexicographic case, the procedure of the statement restricted to the cocircuits vanishing on A defines a valid extension of $\mathcal{M}/a$. Then, by Lemma 1.3 the cocircuit signature of p' defines an extension. The formula $(\mathcal{M} \cup p')/a = (\mathcal{M}/a) \cup p$ holds by construction, which finishes the proof of part (i).

Suppose that p is interior in $(\mathcal{M}/a) \cup p$. Then $p \in \text{conv}_{(\mathcal{M}/a)\cup p}(A)$ for some independent set $A \subseteq E \setminus a$ of $\mathcal{M}/a$ (Lemma 1.1(ii)). But then $A \cup \{a\}$ is independent in $\mathcal{M}$ and $p' \in \text{conv}_{\mathcal{M}\cup p'}(A \cup \{a\})$ (for example, by Lemma 1.1(v)). This proves (ii).

For (iii), suppose that p' is not in general position. Then there is a cocircuit of $\mathcal{M}$ vanishing at p'. By definition of p', this cocircuit vanishes at a. But then the contracted cocircuit vanishes at p and, hence, p is not in general position.

For part (v), observe that an extension p' of $\mathcal{M}$ is a perturbation of the extension parallel to a if and only if $C(p') = C(a)$ for every cocircuit not vanishing at a, and an extension satisfies $(\mathcal{M} \cup p')/a = (\mathcal{M}/a) \cup p$ if and only if $C(p) = C(p')$ for every cocircuit vanishing at a.

In part (vi), the implication from left to right follows from the formula $(\mathcal{M} \cup p')/a = (\mathcal{M}/a) \cup p$. The other implication follows easily from part (iii) of Lemma 1.1: By definition of the cocircuit signature of p', every cocircuit nonnegative on $A \cup \{a\}$ which is not zero at a is positive at p'. For a cocircuit nonnegative on $A \cup \{a\}$ and zero at a, the fact that $p \in \text{conv}_{(\mathcal{M}/a)\cup p}(A)$ implies that the cocircuit (contracted to $\mathcal{M}/a$) is nonnegative at p and hence the (non-contracted) circuit is nonnegative at p'. □

1.3. Euclideanness.

In Definition 7.5.2 of [**11**] a list of several "intersection properties" that an oriented matroid may or may not satisfy is given. Most of the open questions concerning triangulations of oriented matroids have been answered under the assumption that the oriented matroid satisfies one of them. We will be interested in the following two:

DEFINITION 1.11. Let $\mathcal{M}$ be an oriented matroid.

(i) We say that $\mathcal{M}$ has the *Euclidean intersection property* $\mathtt{IP}_3$, or that $\mathcal{M}$ is *Euclidean*, if for every hyperplane H (i.e. flat of corank 1) of $\mathcal{M}$ and every line l (flat of rank 2) of $\mathcal{M}$ there is an extension $\mathcal{M} \cup p$ of $\mathcal{M}$ such that $H \cup \{p\}$ and $l \cup \{p\}$ are respectively a hyperplane and a line of $\mathcal{M} \cup p$.

(ii) We say that $\mathcal{M}$ has the *generalized Euclidean intersection property* $\mathtt{IP}_2$ if for every pair of flats F and G of $\mathcal{M}$ with $\text{rank}(\mathcal{M}) + 1 = \text{rank}(F) + \text{rank}(G)$ there is an extension $\mathcal{M} \cup p$ of $\mathcal{M}$ such that $F \cup \{p\}$ and $G \cup \{p\}$ are flats of of $\mathcal{M} \cup p$ of the same ranks they had in $\mathcal{M}$. Clearly, this is stronger than the condition $\mathtt{IP}_3$.

LEMMA 1.12. *Let $\mathcal{M}$ be an oriented matroid with the generalized Euclidean intersection property $\mathtt{IP}_2$. Let $C = (C^+, C^-)$ be a circuit of $\mathcal{M}$. Then, there is an extension $\mathcal{M} \cup p$ of $\mathcal{M}$ with $p \in relint_{\mathcal{M}\cup p}(C^+)$ and $p \in relint_{\mathcal{M}\cup p}(C^-)$.*

PROOF. There are flats F^+ and F^- containing C^+ and C^- respectively and with $\text{rank}(\mathcal{M}) + 1 = \text{rank}(F^+) + \text{rank}(F^-)$. The property $\mathtt{IP}_3$ guarantees the existence of an extension $\mathcal{M} \cup p$ meeting both flats. In the restriction of $\mathcal{M} \cup p$ to $C^+ \cup C^- \cup \{p\}$ we have that $C^+ \cup \{p\}$ and $C^- \cup \{p\}$ are dependent, i.e. they contain the supports of two circuits. The elimination of p in these circuits must produce a circuit with support contained in $C^+ \cup C^-$, i.e. the original circuit C (or its opposite). Hence, the mentioned circuits of $\mathcal{M} \cup p$ are $(C^+, \{p\})$ and $(C^-, \{p\})$ (or their opposites). Part (iii) of Lemma 1.2 finishes the proof. $\square$

CHAPTER 2

Triangulations of Oriented Matroids

2.1. Definition, characterizations and remarks

We will define a triangulation of an oriented matroid $\mathcal{M}$ of rank r as a collection of r-simplices (i.e. bases) of $\mathcal{M}$ satisfying certain properties. These properties should be the natural translation to oriented matroid terminology of properties characterizing triangulations of point configurations (for the acyclic case) or simplicial fans of vector configurations (for the general case). Candidate properties fall mainly in the following three categories: "covering properties" telling us that the union of the (convex hulls of) simplices of a triangulation covers the convex hull of the configuration; "pseudo-manifold properties" telling us that co-dimension 1 simplices which are not in a facet of $\mathcal{M}$ belong either to 0 or 2 full-dimensional simplices of the triangulation; and "good intersection properties" telling us that the intersection of any two simplices is a face of both. Also, "good intersection properties" are related to "circuit properties" of the simplices, such as "no circuit has its positive and negative parts contained in simplices of a triangulation". The following are translations of these properties to oriented matroid terminology, in several degrees. All of them are satisfied by triangulations of point configurations:

DEFINITION 2.1. Let $\mathcal{M}$ be an oriented matroid of rank r on a set E. Let T be a non-empty collection of r-simplices of $\mathcal{M}$.

- We say that T has the *pseudo-manifold property* if for every $\sigma \in T$, each facet τ of σ is either contained in a facet of $\mathcal{M}$ or there exists another simplex $\sigma' \neq \sigma$ in T with $\tau \subseteq \sigma'$.

  We say that T has the *oriented pseudo-manifold property* if, moreover, for every $(r-1)$-simplex τ contained in at least two r-simplices $\tau \cup a_1$ and $\tau \cup a_2$ of T, the unique (up to sign reversal) cocircuit vanishing on τ has opposite signs at a_1 and a_2. In particular, this implies that τ is not contained in any other element of T.
- We say that T *covers an extension* $\mathcal{M} \cup p$ of $\mathcal{M}$ if $p \in \mathrm{conv}_{\mathcal{M}\cup p}(\sigma)$ for some $\sigma \in T$ (clearly, p has to be an interior extension for this to be possible). We say that T covers the extension *once* if the simplex σ is unique in T.
- We say that two simplices σ_1 and σ_2 of T *intersect properly* if for every one-element extension $\mathcal{M} \cup p$ of $\mathcal{M}$

  $$p \in conv_{\mathcal{M}\cup p}(\sigma_1) \cap conv_{\mathcal{M}\cup p}(\sigma_2) \quad \Longrightarrow \quad p \in conv_{\mathcal{M}\cup p}(\sigma_1 \cap \sigma_2).$$

  We say that the simplices of T intersect properly if this happens for every pair of simplices.
- We say that two simplices σ_1 and σ_2 *overlap on a circuit* $C = (C^+, C^-)$ if $C^+ \subseteq \sigma_1$ and there is an element $a \in C^+$ such that $\underline{C} \setminus \{a\} \subseteq \sigma_2$.

We take as a starting definition of oriented matroid triangulations a slightly modified version of the definition by Billera and Munson in [**9**].

DEFINITION 2.2. Let $\mathcal{M}$ be an oriented matroid of rank r. Let T be a collection of r-simplices of $\mathcal{M}$. We say that T is a *triangulation* if it satisfies the pseudo-manifold property and its simplices intersect properly.

A different definition of triangulations of oriented matroids is given by Anderson in [**1**, Section 2.3]. Roughly speaking, a collection of full-rank simplices is a triangulation if and only if for every (perhaps non-full-rank) simplex τ contained in some simplex of T the link of τ in T is a triangulation of $\mathcal{M}/\tau$. The *link* is a standard concept in piecewise linear topology, but our definition slightly differs from the usual one (compare [**34**]) because we deal with the maximal simplices of a simplicial complex, and not with the whole complex:

DEFINITION 2.3. Let $\mathcal{M}$ be an oriented matroid of rank r. Let T be a collection of r-simplices of $\mathcal{M}$. Let τ be a k-simplex, for some $0 < k \leq r$, contained in some element of T. We call *link* of τ in T the collection of $(r-k)$-simplices $\{\sigma \setminus \tau : \tau \subseteq \sigma, \sigma \in T\}$. We denote it $link_T(\tau)$.

We can now state the main result of Chapter 2, which is the equivalence between several properties, all characterizing triangulations of oriented matroids.

THEOREM 2.4. *Let T be a non-empty collection of r-simplices of an oriented matroid $\mathcal{M}$ of rank r. Then, the following properties are equivalent:*

(a) *The simplices of T intersect properly and T satisfies the pseudo-manifold property (i.e., T is a triangulation of $\mathcal{M}$).*
(b) *The simplices of T intersect properly and T covers every interior extension of $\mathcal{M}$.*
(c) *T satisfies the oriented pseudo-manifold property and covers some interior extension of $\mathcal{M}$ in general position exactly once.*
(d) *T satisfies the oriented pseudo-manifold property and covers all interior extensions of $\mathcal{M}$ in general position exactly once.*
(e) *If $rank(\mathcal{M}) = 1$ then T consists of one simplex if $\mathcal{M}$ is acyclic and two simplices with opposite elements if $\mathcal{M}$ is totally cyclic. If $rank(\mathcal{M}) > 1$ then for every element $a \in E$ the collection $T_a := link_T(a)$ of $(r-1)$-simplices of $\mathcal{M}/a$ is either empty or a triangulation of $\mathcal{M}/a$ and there is an element $a \in E$ such that*
$$\forall \sigma \in T \qquad a \in \sigma \Longleftrightarrow a \in conv_{\mathcal{M}}(\sigma).$$
(f) *T satisfies the pseudo-manifold property and no two simplices of T overlap on a circuit.*
(g) *T satisfies the oriented pseudo-manifold property and for every triangulation T^* of the dual oriented matroid $\mathcal{M}^*$ there is a unique simplex in T whose complement is in T^*.*

REMARKS 2.5. (i) As we have already mentioned, statement (a) of our theorem is essentially the definition of an oriented matroid triangulation by Billera and Munson [**9**] (see also [**11**, Section 9.6]). The differences are that there the oriented matroid $\mathcal{M}$ was assumed to be acyclic and polytopal, and that the original definition included the condition that every element appears in some simplex of the triangulation. This condition is redundant for polytopal oriented matroids. For non-polytopal oriented matroids we prefer to allow triangulations not to use all the elements, as is already done in Section 9.6 of [**11**], because this gives a richer structure to the collection of triangulations of $\mathcal{M}$.

(ii) In the same paper Billera and Munson mention our statement (b) as "a perhaps more obvious parallel to" the usual definition of a triangulation. They observe that (b) implies (a) [**9**, p.520] but discard (b) as a definition because it was not clear to them whether (b) is satisfied for every *lifting triangulation*, while statement (a) was. We will introduce lifting triangulations in Definition 3.4.

(iii) Observe that from (e) it follows by induction the stronger property that for every face τ of a simplex of T the link $link_T(\tau)$ is a triangulation of $\mathcal{M}/\tau$. This is essentially Anderson's definition of a (perhaps partial) triangulation of an oriented matroid (she calls triangulations "partial" if they do not use all the elements). Although she is primarily interested in the totally cyclic case, she proves that in the acyclic and polytopal case statements (e) and (a) are equivalent [**1**, Proposition 5.5].

(iv) Statements (c) and (d) only differ by the word "some" which changes to "all". The combination of the two provides the following simple criterion to decide whether a collection T of r-simplices is a triangulation: first check whether T has the oriented pseudo-manifold property. If yes, choose an arbitrary interior extension in general position and check whether it is covered exactly once by T. If yes, then T is a triangulation, by (c). If not, then T is not a triangulation, by (d). For a lexicographic extension checking this is rather easy.

On the other hand, to check properties (a), (b) or (g) a-priori we need to construct either all the interior extensions of $\mathcal{M}$ in general position or all the triangulations of $\mathcal{M}^*$, which is extremely hard.

Conditions (f) and (e) might be considered at the same level as (c) and (d) for algorithmic purposes, but (f) is specially suitable for the construction of all the triangulations of a fixed oriented matroid: one can iteratively construct all the collections of r-simplices in which no pair overlaps on a circuit and, for the maximal ones, check whether they satisfy the pseudo-manifold property.

(v) Can a triangulation of an oriented matroid have two simplices σ_1 and σ_2 containing respectively the positive and negative part of a circuit $C = (C^+, C^-)$? Lemma 1.12 says that the answer is no if $\mathcal{M}$ satisfies the *Generalized Euclidean intersection property* $\mathtt{IP}_2$, because σ_1 and σ_2 would intersect improperly. Hence, in the presence of $\mathtt{IP}_2$ a collection of full-dimensional simplices is a triangulation if and only if it has the pseudo-manifold property and no pair of simplices contain respectively the positive and negative parts of a circuit. This characterization is used in the realizable case in [**28**, Proposition 2.2]. It is taken as a definition of *circuit admissible triangulation* of an oriented matroid in [**29**].

But we do not know the answer to the question above for general oriented matroids. In Proposition 5.6 we will show two r-simplices in the dual of the non-Euclidean oriented matroid $\mathtt{EMF}(8)$ which contain the positive and negative parts of a circuit and still intersect properly. This example was obtained after Jörg Rambau had shown to the author a similar behaviour in the oriented matroid $R(12)$ of [**33**]. In our example, the two bad simplices do not simultaneously belong to any triangulation (Proposition 5.6). This issue will appear again in Section 2.4, when we discuss what is known about the topological type of oriented matroid triangulations.

2.2. Equivalence of the different characterizations

In this section we prove the equivalence of properties (a) to (f) in Theorem 2.4. The equivalence of properties (a) and (g) is postponed, and appears as part of the statement of Theorem 3.8.

LEMMA 2.6. *Let $\mathcal{M}$ be an oriented matroid of rank r. Let T be a collection of full-rank simplices of $\mathcal{M}$ which intersect properly.*

(i) *Let $\tau = \{a_1, \dots, a_{r-1}\}$ be an $(r-1)$-simplex of $\mathcal{M}$ contained in two different r-simplices $\tau \cup b_1$ and $\tau \cup b_2$ of T. Then the unique (up to sign reversal) cocircuit vanishing on τ has opposite signs at b_1 and b_2. In particular, τ is not in a facet of $\mathcal{M}$ and $\tau \cup b_1$ and $\tau \cup b_2$ are the only simplices of T containing τ.*

(ii) *Let $\mathcal{M} \cup p$ be an extension of $\mathcal{M}$ in general position. Then there is at most one simplex $\sigma \in T$ with $p \in conv_{\mathcal{M}\cup p}(\sigma)$.*

(iii) *No two simplices of T overlap on a circuit.*

PROOF. (i) Let $\tau = \{a_1, \dots, a_{r-1}\}$. Consider the lexicographic extensions of $\mathcal{M}$ by $p_1 = [a_1^+, \dots, a_{r-1}^+, b_1^+]$ and $p_2 = [a_1^+, \dots, a_{r-1}^+, b_2^+]$. The signatures of the two extensions can only differ in the pair of opposite cocircuits vanishing on τ, so our task is to prove that the two extensions do not coincide. But, if they coincide, then $p_1 = p_2 \in conv(\tau \cup b_1) \cap conv(\tau \cup b_2)$ implies that $p_1 \in conv(\tau)$, by the proper intersection property. This is impossible since p_1 and p_2 are in general position by Lemma 1.7(i).

(ii) If $\sigma \neq \sigma'$ are two different simplices of T having p in their convex hull, then $p \in conv_{\mathcal{M}\cup p}(\sigma \cap \sigma')$ violates the general position assumption of p.

(iii) Suppose that two simplices σ_1 and σ_2 overlap on a circuit; that is, there is a circuit $C = (C^+, C^-)$ with $C^+ \subseteq \sigma_1$ and an element $a_1 \in C^+$ such that $\underline{C} \setminus \{a_1\} \subseteq \sigma_2$. Since σ_1 and σ_2 are independent, $C^- \neq \emptyset$ and $a_1 \notin \sigma_2$.

Let $C^+ := \{a_1, \dots, a_k\}$ and consider the lexicographic extension of $\mathcal{M}$ by the element $p := [a_k^+, \dots, a_1^+]$. By part (ii) of Lemma 1.7, $p \in conv_{\mathcal{M}\cup p}(\{a_1, \dots, a_k\}) \subseteq conv_{\mathcal{M}\cup p}(\sigma_1)$. On the other hand, $\sigma_1 \cap \sigma_2 \subseteq \sigma_1 \setminus \{a_1\}$, and the pair of cocircuits vanishing on $\sigma_1 \setminus \{a_1\}$ do not vanish at p. Thus, $p \notin conv_{\mathcal{M}\cup p}(\sigma_1 \cap \sigma_2)$. We will prove that $p \in conv_{\mathcal{M}\cup p}(\sigma_2)$, which violates the proper intersection property of σ_1 and σ_2.

We will use part (v) of Lemma 1.1, since σ_2 is an r-simplex. That is, for each element $a \in \sigma_2$ we consider the cocircuit D vanishing on the facet $\sigma_2 \setminus \{a\}$ of σ_2 and positive at a, and prove that D is non-negative at p. This is clear if $a \in \{a_2, \dots, a_k\} = C^+ \cap \sigma_2$. In every other case $D(p) = D(a_1)$. If $a \in \sigma_2 \setminus \{\underline{C}\}$, then a_1 lies on the flat spanned by $\sigma_2 \setminus \{a\}$ and thus $D(p) = D(a_1) = 0$. If $a \in C^- \subseteq \sigma_2$, then D vanishes on every element of $\underline{C}$ except $a_1 \in C^+$ and $a \in C^-$.

Orthogonality of C and D implies that D has the same sign (possibly zero) at a and a_1, and hence at p. □

For a collection T of geometric d-simplices in $\mathbb{R}^d$, even if it is not a simplicial complex, it makes sense to define the adjacency graph of T as a graph G whose vertices are all the d-simplices and in which two d-simplices are joined by an edge if they have a facet in common. Moreover, given a segment $[a, b]$ in general position with respect to T, one can define the *restriction of G to $[a, b]$* as the subgraph containing only the vertices and edges corresponding to simplices and facets which intersect $[a, b]$. The edges in the graph can naturally be oriented from a to b and the final graph will not have cycles. In the following definition and lemma we try to reproduce this behaviour in the oriented matroid world, the main difference being that we cannot guarantee that cycles do not appear.

DEFINITION 2.7. Let $\mathcal{M}$ be an oriented matroid of rank r.

(a) Let T be a non-empty collection of full rank simplices of $\mathcal{M}$. For any facet τ of a simplex of T which is not contained in a facet of $\mathcal{M}$, let C_τ denote a cocircuit vanishing on τ. We say that T is a *multi-triangulation* if for every such facet τ the following two sets have the same cardinality:

$$\{\tau \cup a \in T : a \in C_\tau^+\}, \qquad \{\tau \cup a \in T : a \in C_\tau^-\}.$$

(b) Let $\mathcal{M}' = \mathcal{M} \cup \{p_1, p_2\}$ be an extension of $\mathcal{M}$ by two interior elements in general position. By this we mean that all the non-zero vectors of $\mathcal{M} \cup \{p_1, p_2\}$ having either p_1 or p_2 (or both) in their support are spanning sets. Let T be a multi-triangulation of $\mathcal{M}$. We call *adjacency graph of T restricted to $[p_1, p_2]$* any directed graph $G_{[p_1,p_2]}$ obtained as follows:
- The nodes of $G_{[p_1,p_2]}$ are the elements σ of T for which $(\{p_1, p_2\}, \sigma)$ is a vector of $\mathcal{M}'$.
- The edges of $G_{[p_1,p_2]}$ arise in the following way: let τ be an $(r-1)$-simplex of $\mathcal{M}$ for which $(\{p_1, p_2\}, \tau)$ is a vector (actually a circuit) of $\mathcal{M}'$. In particular, τ is not in a facet of $\mathcal{M}$. Let $C = (C^+, C^-)$ be the unique cocircuit of $\mathcal{M}'$ vanishing on τ and with $p_1 \in C^-$ and $p_2 \in C^+$. Let $\{\sigma_1^+, \ldots, \sigma_k^+\}$ be the simplices of T containing τ and with $\sigma \setminus \tau \in C^+$, and let $\{\sigma_1^-, \ldots, \sigma_l^-\}$ be the simplices of T containing τ and with $\sigma \setminus \tau \in C^-$. Observe that our notation implicitly introduces an (arbitrary) ordering on the σ_i^+'s and σ_j^-'s. All the simplices σ_i^ϵ satisfy that $(\sigma_i^\epsilon, \{p_1, p_2\})$ is a covector of $\mathcal{M}'$. Hence they are nodes in $G_{[p_1,p_2]}$. Since T is a multi-triangulation, $k = l$. We introduce a directed edge going from σ_i^- to σ_i^+ for each $i = 1, \ldots, k$. We do this for every τ.

Observe that the oriented pseudo-manifold property for a collection T of full-rank simplices of $\mathcal{M}$ can be rephrased as "$\mathcal{M}$ is a multi-triangulation and the cardinality of the sets in the definition of multi-triangulation is always zero or one". In particular, every triangulation is a multi-triangulation as well (if we accept Theorem 2.4, whose proof comes below) and for triangulations the graphs $G_{[p_1,p_2]}$ are uniquely defined: the choice of ordering for the simplices σ_i^ϵ does not appear. We prefer to define the adjacency graphs $G_{[p_1,p_2]}$ in the general context of multi-triangulations as a preparation for Section 3.1 (we will need it in the proof of Proposition 3.1).

From the following lemma it is easy to conclude that the oriented pseudo-manifold property in parts (c), (d) and (f) of Theorem 2.4 can be weakened to "T is a multi-triangulation".

LEMMA 2.8. *Let T be a multi-triangulation of an oriented matroid $\mathcal{M}$ of rank r.*

(i) *There is a certain integer number $m \geq 1$ such that any interior extension of $\mathcal{M}$ in general position is covered by exactly m simplices of T (that is, such that the extension belongs to the convex hull of exactly m simplices).*
(ii) *For any pair of interior extensions p_1 and p_2 in general position, the adjacency graph of T restricted to $[p_1, p_2]$ consists of exactly m "open" components which are either isolated nodes or directed paths and perhaps some "closed" components which are directed cycles.*
(iii) *If $m = 1$, then any two simplices σ and σ' of T are connected by a chain $\sigma = \sigma_0, \ldots, \sigma_k = \sigma'$ in T such that every two consecutive simplices in the chain share a facet.*

PROOF. We start with part (ii) and prove that for any two-element extension $\mathcal{M}' = \mathcal{M} \cup \{p_1, p_2\}$ of $\mathcal{M}$ in general position the connected components of the graph $G_{[p_1,p_2]}$ are either isolated points, or directed paths, or directed cycles. In other words, that $G_{[p_1,p_2]}$ is an oriented 1-manifold with boundary except perhaps for some isolated points. We also claim that the isolated points correspond to r-simplices containing both p_1 and p_2 in the convex hull and that the starting and ending points of the directed paths correspond, respectively, to r-simplices of T having p_1 or p_2 (but not both) in the convex hull. These claims imply that p_1 and p_2 are covered by exactly the same number of simplices of T, namely, by the number m of connected components of $G_{[p_1,p_2]}$ which are not cycles. Since this is true for any choice of p_1 and p_2, we will have proved parts (i) and (ii) (but see the remark at the end of the proof).

If $m = 1$, then the unique "open" component in the graph $G_{[p_1,p_2]}$, where p_1 and p_2 are interior extensions in general position in the convex hulls of simplices σ and σ' of T, produces a chain joining σ to σ'. For example, let p_1 and p_2 be positive lexicographic extensions defined by the elements of σ and σ'. This proves part (iii).

The claims we have made on $G_{[p_1,p_2]}$ follow from the following facts:

(1) if a simplex σ has $p_1 \in conv_{\mathcal{M}'}(\sigma)$ and $p_2 \notin conv_{\mathcal{M}'}(\sigma)$, then it is a node of the graph and there is a unique edge incident to it, which is out-going.

(2) if a simplex σ has $p_1 \notin conv_{\mathcal{M}'}(\sigma)$ and $p_2 \in conv_{\mathcal{M}'}(\sigma)$, then it is a node of the graph and there is a unique edge incident to it, which is in-going.

(3) if a simplex σ has $p_1 \in conv_{\mathcal{M}'}(\sigma)$ and $p_2 \in conv_{\mathcal{M}'}(\sigma)$, then it is an isolated node of the graph.

(4) if a simplex σ has $p_1 \notin conv_{\mathcal{M}'}(\sigma)$ and $p_2 \notin conv_{\mathcal{M}'}(\sigma)$, then either it is not a vertex of the graph, or it is a vertex of the graph with two edges incident to it, one in-going and one out-going.

All the four facts can be easily proved considering the restriction of $\mathcal{M}'$ to $\sigma \cup \{p_1, p_2\}$, which is realizable and uniform (the latter because of the definition of "general position"). In the realized setting, a signed subset $(\{p_1, p_2\}, A)$ is a vector if and only if the relative interiors of $\mathrm{conv}(A)$ and the segment going from p_1 to p_2 intersect.

One remark is in order. Strictly speaking, we have proved that the number m of part (i) is the same for two extensions $\mathcal{M} \cup \{p_1\}$ and $\mathcal{M} \cup \{p_2\}$ only if they are the restrictions of a common two-element extension $\mathcal{M} \cup \{p_1, p_2\}$ of $\mathcal{M}$. We saw in the remark after Lemma 1.7 that not every pair of extensions have this property. If p_1 and p_2 do not have this property, what we can do is consider a lexicographic extension $\mathcal{M} \cup \{p_0\}$ and construct the adjacency graphs for the two-element extensions $\mathcal{M} \cup \{p_1, p_0\}$ and $\mathcal{M} \cup \{p_2, p_0\}$, meaning by these the lexicographic extensions of $\mathcal{M} \cup \{p_1\}$ and $\mathcal{M} \cup \{p_2\}$ with the same expression used in $\mathcal{M} \cup \{p_0\}$. Lemma 1.9 ensures that these two-element extensions restrict to $\mathcal{M} \cup \{p_0\}$ under deletion of p_1 and p_2, respectively. □

As in the realizable case, one would expect that the adjacency graphs $G_{[p_1,p_2]}$ never have cycles, but this is obvious only if the extended oriented matroid $\mathcal{M} \cup \{p_1, p_2\}$ is realizable (more generally, if it is *Euclidean* in the sense of Definition 1.11). For non-Euclidean oriented matroids we neither have a proof that the graph is acyclic nor an example in which it is not.

Showing that if $\mathcal{M} \cup \{p_1, p_2\}$ is Euclidean then the graph $G_{[p_1,p_2]}$ is homeomorphic to either a point or a segment (and a generalization of this to extensions by more than two elements) is a key-step in the proof by Anderson [**1**] of the fact that any triangulation of a Euclidean oriented matroid is PL-homeomorphic to a sphere or ball. Our Lemma 2.8 was inspired by Proposition 3.5 in that paper.

The graph $G_{[p_1,p_2]}$ is also connected to the question of intersection circuits raised in Remark 2.5(v). See Section 2.4 and, in particular, Proposition 2.17.

LEMMA 2.9. *Let T be a collection of simplices of an oriented matroid $\mathcal{M}$ satisfying the oriented pseudo-manifold property. Let τ be a face of one of the simplices of T and let $\mathcal{M} \cup p$ be an extension with $p \in relint_{\mathcal{M}\cup p}(\tau)$. Then,*

(i) *$link_T(\tau)$, considered as a collection of full-rank simplices of $\mathcal{M}/\tau$ has the oriented pseudo-manifold property.*

(ii) *For every perturbation p' of p, interior and in general position, there is an r-simplex σ of T containing τ and with $p' \in conv_{\mathcal{M}\cup p'}(\sigma)$.*

PROOF. (i) The independent sets and bases of the contracted oriented matroid $\mathcal{M}/\tau$ are in bijection with the independent sets and bases of $\mathcal{M}$ containing τ, via the map $\sigma \mapsto \sigma \cup \tau$. Thus, $link_T(\tau)$ is a collection of full-rank simplices in $\mathcal{M}/\tau$. The oriented pseudo-manifold property for T in $\mathcal{M}$ implies the same property for $link_T(\tau)$ in $\mathcal{M}/\tau$.

(ii) The element p' in the contraction $(\mathcal{M} \cup p')/\tau$ is still interior and in general position. By Lemma 2.8, there is a simplex σ in $link_T(\tau)$ with $p' \in conv_{(\mathcal{M}\cup p')/\tau}(\sigma)$. We only need to prove that $p' \in conv_{\mathcal{M}\cup p'}(\sigma \cup \tau)$, since $\sigma \cup \tau \in T$.

We prove this using part (v) of Lemma 1.1: let a be an element in σ, let C be the unique cocircuit C vanishing on $\sigma \cup \tau \setminus \{a\}$ and positive at a. If $a \in \tau$, then C is non-negative on τ, and thus at p (part (iii) of Lemma 1.1) but cannot vanish at p (part (i) of Lemma 1.2). Thus, $C(p) = +1$ and since p' is a perturbation of p, $C(p') = +1$. If $a \notin \tau$, then C vanishes on τ, which implies that C contracts to a cocircuit C_τ of $\mathcal{M}/\tau$ with $C(p') = C_\tau(p'_\tau)$. As p'_τ is in the convex hull of σ, this cocircuit C_τ is nonnegative at p'_τ and C is nonnegative at p'. □

Proof of equivalences (a) to (f) in Theorem 2.4:

(d)⇒(b): Since T covers all interior extensions in general position, it also covers interior extensions not in general position, by part (ii) of Lemma 1.8.

For the proper intersection property, consider an extension $\mathcal{M} \cup p$ and two simplices σ_1 and σ_2 of T with $p \in conv_{\mathcal{M}\cup p}(\sigma_1) \cap conv_{\mathcal{M}\cup p}(\sigma_2)$. Let τ be the minimal face of σ_1 such that $p \in conv_{\mathcal{M}\cup p}(\tau)$. This implies that $p \in relint_{\mathcal{M}\cup p}(\tau)$. We will prove that $\tau \subseteq \sigma_2$, that is, that the simplices intersect properly. Consider the lexicographic perturbation p' of p using the elements of σ_2 with positive signs, so that $p' \in conv_{\mathcal{M}\cup p'}(\sigma_2)$, by part (iv) of Lemma 1.8 and the fact that $p \in conv_{\mathcal{M}\cup p}(\sigma_2)$. Since p' is a perturbation of p into general position, part (ii) of Lemma 2.9 implies that there is a simplex σ' of T containing τ and with $p' \in conv_{\mathcal{M}\cup p'}(\sigma')$. We must have $\sigma' = \sigma_2$ since T covers all interior extensions in general position exactly once.

(b) $\Rightarrow$(a): Let $\tau := \{a_1, \ldots, a_{r-1}\}$ be an $(r-1)$-simplex not contained in a facet of $\mathcal{M}$, and contained in a simplex $\sigma = \tau \cup b$ of T. Since τ is not in a facet, there is an element a_r such that the cocircuit C vanishing on τ has opposite signs at b and a_r. Consider the lexicographic extensions of $\mathcal{M}$ by elements $p := [a_1^+, \ldots, a_{r-1}^+]$ and $p' := [a_1^+, \ldots, a_r^+]$. The extension by p' is interior and in general position and thus there is a simplex σ' in T with $p' \in conv_{\mathcal{M}\cup p'}(\sigma')$. Since p' is a perturbation of p, also $p \in conv_{\mathcal{M}\cup p}(\sigma')$ (part (ii) of Lemma 1.8). But $p \in conv_{\mathcal{M}\cup p}(\sigma)$ as well, and thus $p \in conv_{\mathcal{M}\cup p}(\sigma \cap \sigma')$. The fact that p is in the relative interior of τ implies $\tau \subseteq \sigma'$. This proves the pseudo-manifold property for T.

(a)$\Rightarrow$(c): That the oriented pseudo-manifold property is satisfied is part (i) of Lemma 2.6. For proving that T covers some interior extension in general position exactly once consider the lexicographic extension of $\mathcal{M}$ by $p := [a_1^+, \ldots, a_r^+]$, where $\sigma = \{a_1, \ldots, a_r\}$ is an r-simplex in T. Clearly $p \in conv_{\mathcal{M}\cup p}(\sigma)$. By part (ii) of Lemma 2.6, T covers p only once.

(c) $\Rightarrow$ (d): Straightforward from Lemma 2.8, since the oriented pseudo-manifold property is stronger than the hypothesis in the lemma. This finishes the equivalences (a) $\Leftrightarrow$ (b)$\Leftrightarrow$ (c)$\Leftrightarrow$ (d).

(d) $\Rightarrow$ (e): The case of rank 1 is trivial. Suppose rank$(\mathcal{M}) > 1$. We first claim that for every $a \in E$ contained in some element of T and every $\sigma \in T$ with $a \in \mathrm{conv}_{\mathcal{M}}(\sigma)$ one has $a \in \sigma$. This proves the last assertion in statement (e) since T is not empty.

For the claim, let $\sigma \in T$ with $a \in \mathrm{conv}_{\mathcal{M}}(\sigma)$. If $a \notin \sigma$ then there is a subset $\tau \subseteq \sigma$ such that $(\{a\}, \tau)$ is a circuit. Let p be the extension of $\mathcal{M}$ parallel to a, and let p' be any perturbation of p interior and in general position. By part (ii) of Lemma 2.9, applied once with $\{a\}$ and once with τ as the face of a simplex of T, there are simplices σ_1 and σ_2 in T covering p' and containing respectively a and τ. These two simplices must be different, since $\tau \cup \{a\}$ is dependent. But this is impossible since T satisfies (d). This proves that $a \in \sigma$.

Now we prove that for every $a \in E$, $\mathrm{link}_T(a)$ is either empty or a triangulation of $\mathcal{M}/a$. If it is not empty, it has the oriented pseudo-manifold property (considered in $\mathcal{M}/a$) by part (i) of Lemma 2.9. We will prove that $\mathrm{link}_T(a)$ satisfies characterization (c) of triangulations.

Let p' be any perturbation interior and in general position of the extension of $\mathcal{M}$ parallel to a. Let p denote the extension of $\mathcal{M}/a$ defined by $(\mathcal{M}/a)\cup p = (\mathcal{M}\cup p')/a$. Every $\sigma \in T$ with $p' \in \mathrm{conv}_{\mathcal{M}\cup p'}(\sigma)$ has also $a \in \mathrm{conv}_{\mathcal{M}\cup p'}(\sigma)$ (part (ii) of Lemma 1.8) and, hence, by what we proved above, it has $a \in \sigma$. By (d), exactly one such σ exists. Clearly one also has $p \in \mathrm{conv}_{(\mathcal{M}\cup p)/a}(\sigma \setminus \{a\})$.

We have to prove that no other $\sigma_a \in \text{link}_T(a)$ has $p' \in \text{conv}_{(\mathcal{M}\cup p')/a}(\sigma_a)$. In other words, that if $p' \in \text{conv}_{(\mathcal{M}\cup p')/a}(\sigma_a)$ then the pair of cocircuits of $\mathcal{M}$ vanishing on σ_a have the same sign at a and p' (and hence $p' \in \text{conv}_{\mathcal{M}\cup p'}(\sigma_a \cup \{a\})$). But this is trivial since those cocircuits cannot vanish at a and since p' is a perturbation of the extension parallel to a.

(e)⇒(c): The fact that the link of every element is either empty or a triangulation and that triangulations satisfy (e) recursively implies that the link of every simplex τ is either empty or a triangulation of $\mathcal{M}/\tau$. This property applied to the simplices of rank $rank(\mathcal{M}) - 1$ is stronger than the oriented pseudo-manifold property.

In order to find an extension which is covered exactly once, let a be an element such that $a \in conv_M(\sigma)$ and $\sigma \in T$ implies that $a \in \sigma$. Let $\sigma := \{a, a_2, \ldots, a_r\}$ be an r-simplex of T containing a (it exists since a, considered as an extension, has to be covered by a simplex of T having a as a vertex). Consider the lexicographic extension by the element $p := [a^+, a_2^+, \ldots, a_r^+]$. We have that $p \in conv_{\mathcal{M}\cup p}(\sigma)$ and will show that σ is the only simplex of T with this property.

For this we consider the contraction $(\mathcal{M} \cup p)/a$ of $\mathcal{M} \cup p$, which is an interior extension in general position of $\mathcal{M}/a$, and denote it $(\mathcal{M}/a) \cup p_a$. The extensions $(\mathcal{M} \cup p)/a$ and $(\mathcal{M}/a) \cup p_a$ are in the conditions of Lemma 1.10. The simplex $\{a_2, \ldots, a_r\}$ is the unique element of T_a having p_a in its convex hull, since T_a is a triangulation. By part (vi) of Lemma 1.10 $\sigma = \{a, a_2, \ldots, a_r\}$ is the unique simplex of T containing a and which has p in its convex hull. On the other hand, simplices of T not containing a do not have a in their convex hull and, thus, do not have p. Hence, T covers p exactly once.

(a)⇒(f) Follows from part (iii) of Lemma 2.6.

(f)⇒(e) The case of rank 1 is trivial. For the general case, let a be an arbitrary vertex of a simplex of T. If $a \in conv_{\mathcal{M}}(\sigma)$ for a simplex $\sigma \in T$ with $a \notin \sigma$, then there is a circuit of $\mathcal{M}$ of the form $(\{a\}, B)$ with $B \subseteq A$. That is, σ and any simplex of T having a as a vertex overlap on the circuit $(\{a\}, B)$. This proves the last part of (e).

We will now prove that the link of every vertex of T satisfies (f). Inductively we assume that (f) and (a) are equivalent in rank lower than $rank(\mathcal{M})$, which implies that the link of every vertex in T is a triangulation.

Thus, let $a \in E$ be a vertex of T. The pseudo-manifold property for $link_T(a)$ follows from the pseudo-manifold property of T in the same way as we proved the oriented pseudo-manifold property in part (i) of Lemma 2.9. Now suppose that two simplices τ_1 and τ_2 of $link_T(a)$ overlap on a circuit $C = (C^+, C^-)$ of $\mathcal{M}/a$. That is, $C^+ \subseteq \tau_1$ and there is an element a_1 in C^+ such that $\underline{C} \setminus \{a_1\} \subseteq \tau_2$. If (C^+, C^-) is a circuit of $\mathcal{M}/a$, then one of (C^+, C^-), $(C^+ \cup \{a\}, C^-)$ and $(C^+, C^- \cup \{a\})$ is a circuit in $\mathcal{M}$. In the three cases we have that $\tau_1 \cup \{a\}$ and $\tau_2 \cup \{a\}$ overlap in that circuit. □

2.3. Some properties of triangulations

Here we prove several properties of triangulations which either are interesting by themselves or will be used later on. Given two collections A and B of subsets of two disjoint sets E and F respectively, the *join* $A \cdot B$ denotes the following collection of subsets of $E \cup F$:

$$A \cdot B := \{\tau \cup \sigma : \tau \in A, \sigma \in B\}.$$

We will use $A \cdot b$ as an abbreviation for $A \cdot \{\{b\}\}$.

PROPOSITION 2.10. *Let $\mathcal{M}$ be an oriented matroid of rank r on a set E and let a be any element of E. Let T be a triangulation of the restricted oriented matroid $\mathcal{M}(E \setminus a)$.*

(i) *If $a \in conv_{\mathcal{M}}(E \setminus \{a\})$, then T is a triangulation of $\mathcal{M}$.*
(ii) *If $a \notin conv_{\mathcal{M}}(E \setminus a)$ (i.e., a is a vertex of $\mathcal{M}$), let T_a be the collection of facets of simplices of T which "are visible" from a. More precisely, an $(r-1)$-simplex τ of $\mathcal{M}$ is in T_a if and only if it is contained in a simplex of T and there is a cocircuit of $\mathcal{M}$ which is zero on τ, positive at a and non-positive on $E \setminus \{a\}$. Then, $T \cup (T_a \cdot a)$ is a triangulation of $\mathcal{M}$.*
(iii) *In the conditions of (ii), T_a is a triangulation of $\mathcal{M}/a$.*

Moreover, in parts (i) and (ii) the triangulation of $\mathcal{M}$ exhibited is the only triangulation of $\mathcal{M}$ which extends T.

PROOF. (i) This is obvious from characterization (c) in Theorem 2.4.

(ii) We will prove that $T' := T \cup (T_a \cdot a)$ satisfies characterization (c) of Theorem 2.4. We first prove that T' covers some interior extension exactly once. Consider an interior extension $\mathcal{M} \cup p$ with $p \in conv_{\mathcal{M}\cup p}(E \setminus a)$. Since T is a triangulation of $\mathcal{M}(E \setminus a)$ the simplices of T cover p exactly once. In the other hand, the simplices of the form $\tau \cup a$ with $\tau \in T_a$ do not cover p, since there is a cocircuit vanishing on τ and with opposite signs at p and a (compare Lemma 1.1(v)).

Secondly we prove that T' satisfies the oriented pseudo-manifold property: for those $(r-1)$-simplices which are interior in $\mathcal{M}(E \setminus a)$ this is clear, from the oriented pseudo-manifold property of T. For an $(r-1)$-simplex $\tau = \{a_1, \dots, a_{r-1}\}$ in T_a, let a_r be any negative element of the cocircuit positive at a and which vanishes on τ. There is at least one r-simplex of T containing τ, by definition of T_a, and at most one because any such simplex covers the lexicographic extension $p = [a_1^+, \dots, a_{r-1}^+, a_r^-]$ of $\mathcal{M}(E \setminus a)$, which is interior and in general position. This simplex and $\tau \cup \{a\}$ are the only simplices in $T \cup (T_a \cdot a)$ containing τ and they are in the conditions of the oriented pseudo-manifold property by definition of visible.

We finally have to deal with the $(r-1)$-simplices of T' which are interior and use the element a. These simplices are of the form $\rho \cup a$, where ρ is an $(r-2)$-simplex in a facet of $\mathcal{M}(E \setminus a)$ such that $\rho \cup b_1$ is a boundary $(r-1)$-simplex of T visible from a for some $b_1 \in E \setminus a$.

Consider the link $L := link_T(\rho)$. By characterization (e) of Theorem 2.4, L is a triangulation of the oriented matroid $\mathcal{M}(E \setminus a)/\rho$, which has rank 2 and is not totally cyclic. Rank 2 easily implies the following properties: L has exactly two boundary rank-1 simplices. One is $\{b_1\}$ for the afore mentioned b_1 and we let the other be $\{b_2\}$ (i.e. ρ is contained in exactly two boundary $(r-1)$-simplices $\tau_1 = \rho \cup \{b_1\}$ and $\tau_2 = \rho \cup \{b_2\}$ of the triangulation T). Since a is interior in $\mathcal{M}/\rho$ but an exterior extension of $\mathcal{M}(E \setminus a)/\rho$ (because $\rho \cup \{a\}$ is interior in $\mathcal{M}$ but a is the only positive element in one of the two cocircuits vanishing on $\rho \cup b_1$) both b_1 and b_2 are visible from a and they lie in opposite sides of the cocircuits (of $\mathcal{M}/\rho$) vanishing at a. We conclude that the link of $\rho \cup a$ in $T_a \cdot a$ consists exactly of the two elements b_1 and b_2, and that they lie on opposite sides of the cocircuits vanishing on $\rho \cup a$. I.e. we conclude the oriented pseudo-manifold property for $\rho \cup a$.

(iii) Since $T_a = link_{T \cup (T_a \cdot a)}(a)$, this follows from part (ii) and characterization (e) in Theorem 2.4.

The final sentence is trivial: both in parts (i) and (ii) it is obvious that any triangulation of $\mathcal{M}$ extending T will contain the one we state. But two triangulations of an oriented matroid cannot be one properly contained in the other, for example by characterization (d) of Theorem 2.4. □

Proposition 2.10 provides a natural way of iteratively constructing a triangulation of an oriented matroid $\mathcal{M}$. Let k be the corank of $\mathcal{M}$ and let $\sigma^* = \{a_1, \ldots, a_k\}$ be any basis of the dual oriented matroid $\mathcal{M}^*$, whose elements we consider ordered by their labels. Let $\sigma = E \setminus \sigma^*$, which is a basis of $\mathcal{M}$. Then, $T_0 := \{\sigma\}$ is a triangulation of $\mathcal{M}(\sigma)$ and for each $i = 1, \ldots, k$ we let T_i be the unique triangulation of $\mathcal{M}(\sigma \cup \{a_1, \ldots, a_i\})$ which extends T_{i-1}. The triangulation T_k of $\mathcal{M}$ obtained in this way is well-known in the realized case and called the *placing triangulation* for that specific order of the elements (see [**25**]). It is also called *pushing triangulation* for reasons which will become apparent in Remark 4.4. Our next result shows that this triangulation is a particular case of the *lexicographic triangulations* that will be introduced in Definition 3.4. In the statement and in the proof, $\Delta(\mathcal{M})$ denotes the collection of all full-rank simplices (bases) of an oriented matroid $\mathcal{M}$.

PROPOSITION 2.11. *Let $\mathcal{M}$ be an oriented matroid of rank r and corank k, with dual $\mathcal{M}^*$. Let σ be a basis of $\mathcal{M}$ and let its complement be $\sigma^* = \{a_1, a_2, \ldots, a_k\}$, with its elements ordered as shown by their labels. Let $\mathcal{M}^* \cup p$ be the positive lexicographic extension of $\mathcal{M}^*$ with $p := [a_k^+, \ldots, a_2^+, a_1^+]$.*

Then, the placing triangulation of $\mathcal{M}$ associated with the given ordering of the elements of σ^*, i.e. the triangulation obtained by starting with σ and then placing the points of σ^* in the order of their labels, equals

$$\{\tau \in \Delta(\mathcal{M}) : p \in conv_{\mathcal{M}^* \cup p}(E \setminus \tau)\}.$$

PROOF. For any interior extension p in general position of the dual $\mathcal{M}^*$, the collection $T_p := \{\tau \in \Delta(\mathcal{M}) : p \in \mathrm{conv}_{\mathcal{M}^* \cup p}(E \setminus \tau)\}$ is a triangulation of $\mathcal{M}$ (this will be proved in Corollary 3.3). In the case of the lexicographic extension $p := [a_k^+, \ldots, a_2^+, a_1^+]$, Lemma 1.10 (specially parts (iv) and (vi)) implies that T_p extends the triangulation $T_{p'} := \{\tau \in \Delta(\mathcal{M} \setminus a_k) : p' \in \mathrm{conv}_{(\mathcal{M}^* \cup p')/a_k}(E \setminus (\tau \cup \{a_k\}))\}$ of $\mathcal{M} \setminus a_k$ given by the lexicographic extension $p' = [a_{k-1}^+, \ldots, a_2^+, a_1^+]$ of $\mathcal{M}^*/a_k$.

By induction on k, $T_{p'}$ is the placing triangulation of $\mathcal{M} \setminus a_k$. Then, uniqueness in Proposition 2.10 implies that T_p is the placing triangulation as well. □

We now consider the following notion of *restriction of a triangulation to a face*. Let T be a triangulation of an oriented matroid $\mathcal{M}$ and let F be a face of $\mathcal{M}$ of rank k. We will call restriction of T to F the following collection of full-rank-simplices of $\mathcal{M}(F)$:

$$\{\tau : rank(\tau) = k, \quad \tau \subseteq F, \quad \exists \sigma \in T \text{ with } \tau \subseteq \sigma\}.$$

COROLLARY 2.12. *Let T be a triangulation of an oriented matroid $\mathcal{M}$ and F be a face of $\mathcal{M}$ of rank k. Then, the restriction of T to F is a triangulation of $\mathcal{M}(F)$.*

PROOF. Using recursion we only need to prove the case of F being a facet, i.e., $rank(\mathcal{M}) = k + 1$. In this case let $\{a_1, \ldots, a_k\}$ be a k-simplex in F and let $b \notin F$ be an element of $\mathcal{M}$. Then the lexicographic extension by $p := [a_1^+, \ldots, a_k^+, b^-]$ is exterior to $\mathcal{M}$. Moreover, the triangulation T_p of $(\mathcal{M} \cup p)/p$ of part (iii) of Proposition 2.10 is precisely the collection of simplices T_F. Observe that although

$(\mathcal{M} \cup p)/p$ and $\mathcal{M}(F)$ are different oriented matroids, the first one is an extension of the second one by some interior elements. This implies that $T_p = T_F$ is a triangulation of $\mathcal{M}(F)$ as well, as stated in Proposition 2.10. □

PROPOSITION 2.13. *Let $\mathcal{M}$ be an oriented matroid of corank 1 on a set E. Let $C = (C^+, C^-)$ be one of the two (opposite) circuits of $\mathcal{M}$. Then,*

- *If both parts of C are non-empty (i.e., if $\mathcal{M}$ is acyclic) the only triangulations of $\mathcal{M}$ are*
$$T^+ := \{E \setminus \{e\} : e \in C^+\}$$
and
$$T^- := \{E \setminus \{e\} : e \in C^-\}.$$
- *Otherwise, the only triangulation of $\mathcal{M}$ is*
$$T := \{E \setminus \{e\} : e \in \underline{C}\}.$$

PROOF. The oriented matroid $\mathcal{M}$ has one more element than its rank. Actually, its maximal simplices (bases) are the subsets $\underline{C} \setminus \{e\}$, for $e \in \underline{C}$. In case (ii) this implies that any triangulation of $\mathcal{M}$ is contained in T. In case (i), the fact that no two simplices of a triangulation overlap on a circuit (characterization (f) in Theorem 2.4) implies that every triangulation is contained in either T^+ or T^-.

Since no two triangulations of an oriented matroid can be contained in one another, we will have finished if we prove that T, T^+ and T^- are in fact triangulations. This is easy to verify; it will also be a trivial consequence of Corollary 3.3 in the next section, since the dual $\mathcal{M}^*$ has rank 1. □

2.4. Topology of triangulations

For any triangulation T of an oriented matroid $\mathcal{M}$ we consider the simplicial complex $\mathcal{P}(T)$ of which T is the collection of maximal simplices. If $\mathcal{M}$ is realizable of rank r, then $\mathcal{P}(T)$ is PL-homeomorphic to an $(r-1)$-sphere or to an $(r-1)$-ball, depending on whether $\mathcal{M}$ is totally cyclic or not. Probably the central open problem in the theory of oriented matroid triangulations is whether this is also the case for non-realizable oriented matroids. Here we give a partial answer, following mostly the work of Laura Anderson [**1**].

PROPOSITION 2.14. *Let T be a triangulation of an oriented matroid $\mathcal{M}$ of rank r. Let $\mathcal{P}(T)$ be the simplicial complex induced by T.*

(i) *If $\mathcal{M}$ is totally cyclic, then $\mathcal{P}(T)$ is an orientable, strongly connected, pseudo-manifold of dimension $r-1$ without boundary.*
(ii) *Otherwise, $\mathcal{P}(T)$ is an orientable, strongly connected, pseudo-manifold of rank $r-1$ whose boundary is a triangulation of a totally cyclic oriented matroid of rank $r-1$. If $\mathcal{M}$ is uniform, then the boundary is a PL-sphere of dimension $r-2$.*

PROOF. The oriented pseudo-manifold property of a triangulation T implies that $\mathcal{P}(T)$ is a pseudo-manifold whose boundary is made of $(r-1)$-simplices all lying in proper faces of $\mathcal{M}$ and that the chirotope of $\mathcal{M}$ induces a coherent orientation on $\mathcal{P}(T)$. Part (ii) of Lemma 2.8 has the consequence that $\mathcal{P}(T)$ is strongly connected (this appears also in [**1**, Proposition 4.2]). This proves the statement, except for the assertions concerning the boundary of $\mathcal{P}(T)$.

If $\mathcal{M}$ is totally cyclic, the boundary is clearly empty. If not, let p be an exterior extension of $\mathcal{M}$, meaning that $p \notin \mathrm{conv}_{\mathcal{M}\cup p}(E)$, where E is the ground set of $\mathcal{M}$. $\mathcal{M} \cup p$ is a totally cyclic oriented matroid. Applying the procedure of Proposition 2.10 we get a triangulation $T' = T \cup (T_p \cdot p)$ of $\mathcal{M} \cup p$ which extends T. The boundary of $\mathcal{P}(T)$ clearly coincides with $\mathrm{link}_{T'}(p)$, which is a triangulation of $(\mathcal{M} \cup p)/p$.

If $\mathcal{M}$ is uniform or, more generally, if every proper face of $\mathcal{M}$ is a simplex, then the boundary of $\mathcal{P}(T)$ is precisely the Las Vergnas face lattice of $\mathcal{M}$, which is a PL-sphere of dimension $r - 2$ by [**11**, Proposition 9.1.1]. (Remark: that result assumes $\mathcal{M}$ to be acyclic, but can be adapted to the non-totally cyclic case, which is more general, without difficulty). □

PROPOSITION 2.15. *Let r be a natural number. The following statements are equivalent:*

(a) *For every triangulation T of every rank $r + 1$ oriented matroid, $\mathcal{P}(T)$ is an r-manifold (possibly with boundary).*
(b) *For every triangulation T of every rank r oriented matroid, $\mathcal{P}(T)$ is an $(r - 1)$-sphere if $\mathcal{M}$ is totally cyclic and an $(r - 1)$-ball otherwise.*
(c) *For every triangulation T of every totally cyclic rank r oriented matroid, $\mathcal{P}(T)$ is an $(r - 1)$-sphere.*

PROOF. The implications from (b) to both (a) and (c) are obvious. The implication from (a) to (b) is also obvious, taking into account that every triangulation T of rank r appears as a link in a triangulation of rank $r + 1$, namely the cone of T with apex in a coloop.

For the implication from (c) to (b), observe that any non-totally cyclic oriented matroid $\mathcal{M}$ has a totally cyclic extension $\mathcal{M} \cup p$ (take as p the opposite of any relative interior extension). Any triangulation T of $\mathcal{M}$ extends to a triangulation T' of $\mathcal{M} \cup p$ with the placing procedure of Proposition 2.10. $\mathcal{P}(T)$ is the antistar of p in the simplicial complex $\mathcal{P}(T')$ and, thus, it is an $(r - 1)$-ball. □

Any of the conditions in the statement of Proposition 2.15 is easy to check for the class of *lifting triangulations* defined in [**11**, Section 9.6], using that the proper faces of an acyclic oriented matroid of rank r form a poset isomorphic to a cell decomposition of an $(r - 2)$-sphere [**11**, Proposition 9.1.1]. Lifting triangulations will be of central importance in the second half of this paper (see Definitions 3.4 and 4.1). Anderson [**1**] has shown that the conditions hold when restricted to Euclidean oriented matroids. Observe that all oriented matroids of rank up to 3 are Euclidean.

THEOREM 2.16 (Anderson). *All triangulations of totally cyclic Euclidean oriented matroids of rank r are spheres of dimension $r - 1$.*

SKETCH OF PROOF (see [**1**] for details). Let $\mathcal{M}$ be a totally cyclic oriented matroid. Then:

(i) $\mathcal{M}$ has triangulations which are spheres; for example, lifting triangulations. Hence, we only need to prove that any two triangulations T_1 and T_2 of $\mathcal{M}$ are PL-homeomorphic.
(ii) Without loss of generality we assume that T_1 and T_2 do not use any common element of $\mathcal{M}$. To obtain this, "double" each element of $\mathcal{M}$ (i.e., extend by an element parallel to it) and consider one copy of the element as used in T_1 and the other in T_2.

(iii) Consider the collection $R_{\mathcal{M}}(T_1, T_2)$ of vectors of $\mathcal{M}$ with positive part in $\mathcal{P}(T_1)$ and negative part in $\mathcal{P}(T_2)$. This collection of vectors can be given a topology in two equivalent ways: either as the order complex of its natural poset structure or as a subcomplex of the topological representation of the dual $\mathcal{M}^*$ as a cell decomposition of a sphere. If $\mathcal{M}$ is a realized oriented matroid then $R_{\mathcal{M}}(T_1, T_2)$ coincides with the face poset of the common refinement of T_1 and T_2 that one gets by superimposing the two triangulations.

(iv) Using Euclideanness of $\mathcal{M}$, prove that $R_{\mathcal{M}}(T_1, T_2)$ is indeed a PL-refinement of both $\mathcal{P}(T_1)$ an $\mathcal{P}(T_2)$. This is the hard part of the proof. □

Part of step (iv) above consists in proving that the adjacency graphs $G_{[p_1,p_2]}$ of Definition 2.7 and Lemma 2.8 cannot have cycles, for triangulations of Euclidean oriented matroids. Our following statement, due to Rambau, shows that this property is in turn related to the question of "circuit intersections" between simplices of a triangulation, in the sense of Remark 2.5(v). Remember that:

- In Definition 2.8, from a triangulation T of $\mathcal{M}$ and an extension $\mathcal{M} \cup \{p_1, p_2\}$ by two interior elements in general position we constructed the adjacency graph $G_{[p_1,p_2]}$ whose vertices and edges were respectively the full-rank and corank 1 simplices τ in T such that $(\tau, \{p_1, p_2\})$ is a vector in $\mathcal{M} \cup \{p_1, p_2\}$. We proved in Lemma 2.8 that the graph consists of a distinguished "open" component homeomorphic to a point or a segment and, perhaps, some closed components homeomorphic to cycles.
- In Remark 2.5(v) we asked whether a circuit $C = (C^+, C^-)$ can exist with both C^+ and C^- in $\mathcal{P}(T)$. If this happens, let us say that the circuit C is an *intersection circuit* of T. Part (g) of Theorem 2.4 implies that C^+ and C^- have at least two elements each if C is an intersection circuit.

PROPOSITION 2.17 (Rambau). *Let T be a triangulation of an oriented matroid $\mathcal{M}$:*

(i) *If the graph $G_{[p_1,p_2]}$ has a cycle for some pair of extensions, then T has an intersection circuit.*

(ii) *Conversely, if T has an intersection circuit and one of C^+ or C^- has only 2 elements, then the graph $G_{[p_1,p_2]}$ constructed with certain perturbations of those two elements has a cycle.*

PROOF. Part (i) is the main result in [**29**]. Let us prove (ii).

Let (C^+, C^-) be an intersection circuit of a triangulation T, so that both C^+ and C^- are contained in simplices of T. Suppose further that C^+ has only two elements.

Let $C^+ = \{a_1, a_2\}$ and $C^- = \{b_1, \dots, b_k\}$ $(k \leq r-1)$. Let $\tau = \{b_1, \dots, b_{r-1}\}$ be a corank 1 simplex of T containing C^- and such that $\tau \cup \{a_1\}$ is a basis. This can be obtained starting with any full-rank simplex σ of T containing C^- and extracting from $\sigma \cup \{a_1\}$ a basis which extends the independent set $C^- \cup \{a_1\}$. We will have that $\tau \cup \{a_2\}$ is a basis too, since the unique circuit with support in $\tau \cup \{a_1, a_2\}$ is $(\{a_1, a_2\}, C^-)$.

Consider the positive lexicographic extension of $\mathcal{M}$ by two elements $p_1 := [a_1^+, a_2^+, b_1^+, \dots, b_{r-1}^+]$ and $p_2 := [a_2^+, a_1^+, b_1^+, \dots, b_{r-1}^+]$. More precisely, first extend $\mathcal{M}$ by p_1 and then $\mathcal{M} \cup \{p_1\}$ by p_2. Observe that the lexicographic expressions contain redundancy, since the elements in them are not independent. Let $\mathcal{M}' = \mathcal{M} \cup \{p_1, p_2\}$.

We first look at the contraction of $\mathcal{M}$ and $\mathcal{M}'$ at $\{a_1, a_2\}$. The extensions p_1 and p_2 become two lexicographic extensions in general position obtained *with the same* lexicographic expression. In particular, both of them lie in the relative interior of a unique and the same simplex of the triangulation $\mathrm{link}_T(\{a_1, a_2\})$. Let σ be this simplex. Then, $\sigma \cup \{a_1, a_2\}$ is a simplex of T containing both p_1 and p_2 in its relative interior, by part (vi) of Lemma 1.10. In particular, $\sigma \cup \{a_1, a_2\}$ represents an isolated node in the graph $G_{[p_1,p_2]}$.

On the other hand, Lemma 2.8, together with characterrizations (c) or (d) of Theorem 2.4, implies that all but one of the components of $G_{[p_1,p_2]}$ are cycles. Hence, we just need to prove that $\sigma \cup \{a_1, a_2\}$ is not the only node in $G_{[p_1,p_2]}$.

The oriented matroid $\mathcal{M}(\tau \cup \{a_1, a_2\})$ is realizable and has (C^+, C^-) as its unique circuit. In particular, in any realization of it the segment $\mathrm{conv}(\{a_1, a_2\})$ and the simplex $\mathrm{conv}(\tau)$ intersect in an interior point of the former. The restriction of $\mathcal{M}'$ to $\tau \cup \{a_1, a_2, p_1, p_2\}$ is a lexicographic two-element extension of $\mathcal{M}(\tau \cup \{a_1, a_2\})$ obtained by perturbing the points a_1 and a_2 in a certain way "towards the interior". In a suitable realization, we will have p_1 and p_2 in the relative interior and "arbitrarily close" to a_1 and a_2 respectively. The description of $\mathcal{M}(\tau \cup \{a_1, a_2\})$ above implies that (in the realization) the segment $\mathrm{conv}(\{p_1, p_2\})$ and the simplex $\mathrm{conv}(\tau)$ intersect in a point in the relative interior of both. I.e., $(\{p_1, p_2\}, \tau)$ is a vector in $\mathcal{M}'(\tau \cup \{a_1, a_2, p_1, p_2\})$ and, hence, also in $\mathcal{M}'$. In particular, τ is an edge in the graph $G_{[p_1,p_2]}$. □

To close the circle of concepts, there is a simple reason why Anderson's proof of Theorem 2.16 has little hope of working for triangulations that contain intersection circuits, in case they exist. Let T be such a triangulation and let's try to apply the ideas in the proof of Theorem 2.16 with $T_1 = T_2 = T$. We get that the poset $R_{\mathcal{M}}(T, T)$ indeed has a part homeomorphic to $\mathcal{P}(T)$, consisting of vectors whose positive and negative parts are parallel copies of the different simplices in $\mathcal{P}(T)$. But the intersection circuits provide extra cells in $R_{\mathcal{M}}(T, T)$. Hence, $R_{\mathcal{M}}(T, T)$ is no longer a PL-refinement of $\mathcal{P}(T)$.

⋆

Let us finish with a description of the situation in rank 4. This is the first open case, since every rank 3 oriented matroid is Euclidean. We know by Propositions 2.14 and 2.15 that rank 4 triangulations are orientable connected manifolds, without boundary if $\mathcal{M}$ is totally cyclic and with a boundary homeomorphic to S^2 otherwise. Knowing that they are simply connected would imply that they are homotopy equivalent to 3-spheres or 3-balls (and "homeomorphic to", modulo Poincaré conjecture...) Studying the fundamental group of triangulations seems to be a crucial step in resolving the topology problem. This study essentially boils down to have an analogue of the graph $G_{[p_1,p_2]}$ and, more importantly, of Lemma 2.8, for extensions by three elements.

As for the adjacency graphs and intersection circuits, observe that circuits in rank 4 have at most 5 elements and, hence, at most two elements in one of the sides. Proposition 2.17 implies that in rank 4 the existence of intersection circuits is equivalent to the existence of extensions for which the adjacency graph have cycles.

CHAPTER 3

Duality between Triangulations and Extensions

3.1. Circuit, cocircuit, extension and triangulation vectors

The goal of this section is to prove that every interior extension in general position of an oriented matroid $\mathcal{M}$ has associated a triangulation of the dual oriented matroid $\mathcal{M}^*$. The triangulations which can be obtained in this way will be called *lifting triangulations* of $\mathcal{M}^*$ (Definition 3.4) and in a certain sense are the analogue in oriented matroid terms of the *regular triangulations* (see Examples 5.1) of a point configuration. A different, more geometric, definition of lifting triangulations appears in [**11**, Section 9.6]. In Section 4.1 we will prove the equivalence of the two definitions. Definition 3.4 makes more explicit the importance of duality in the context of triangulations.

The use of lifting triangulations will allow us to extend to the non-realizable case most of the results in Sections 2 and 5 of [**14**]. In particular, the following notations come from Section 5 in that paper.

Let $\mathcal{M}$ be an oriented matroid of rank r. Let $\Delta(\mathcal{M})$ denote the collection of all bases (r-simplices) of $\mathcal{M}$. Let e_σ denote the standard basis vector of $\mathbb{R}^{\Delta(\mathcal{M})}$ corresponding to an r-simplex σ of $\mathcal{M}$. For any triangulation $T \subseteq \Delta(\mathcal{M})$ we consider its characteristic vector $v_T \in \mathbb{R}^{\Delta(\mathcal{M})}$, which has coordinates $(v_T)_\sigma = 1$ if $\sigma \in T$ and $(v_T)_\sigma = 0$ if $\sigma \notin T$.

Let τ be an $(r-1)$-simplex of $\mathcal{M}$. Let $C = (C^+, C^-)$ be the unique (up to sign reversal) cocircuit vanishing on τ. We define the *cocircuit vector* $Co_\tau \in \{-1, 0, 1\}^{\Delta(\mathcal{M})}$ by

$$Co_\tau \quad := \quad \sum_{i \in C^+} e_{\tau \cup i} \quad - \quad \sum_{j \in C^-} e_{\tau \cup j}.$$

We say that a cocircuit vector is *interior* if both $+1$ and -1 appear among the coordinates of Co_τ (i.e., if τ is not in a facet of $\mathcal{M}$). Let $Co(\mathcal{M})$ denote the collection of all cocircuit vectors Co_τ, where τ runs over all $(r-1)$-simplices of $\mathcal{M}$. We denote by $Co_{int}(\mathcal{M})$ the set of interior cocircuit vectors, i.e of cocircuit vectors which have positive and negative entries. $\mathcal{M}$ is totally cyclic if and only if $Co(\mathcal{M}) = Co_{int}(\mathcal{M})$.

Dually, let ρ be a spanning $(r+1)$-subset of $\mathcal{M}$. Then ρ contains a unique (up to sign reversal) signed circuit $C = (C^+, C^-)$ of $\mathcal{M}$. We define the *circuit vector* $Ci_\rho \in \{-1, 0, 1\}^{\Delta(\mathcal{M})}$ by

$$Ci_\rho \quad := \quad \sum_{a \in C^-} e_{\rho \setminus \{a\}} \quad - \quad \sum_{a \in C^+} e_{\rho \setminus \{a\}}.$$

We say that Ci_ρ is an *acyclic* circuit vector if both $+1$ and -1 appear among the coordinates of Ci_ρ (i.e., if the restriction $\mathcal{M}(\rho)$ is acyclic). Let $Ci(\mathcal{M})$ denote the set of all circuit vectors and $Ci_{ac}(\mathcal{M})$ the subset of acyclic circuit vectors. $\mathcal{M}$ is acyclic if and only if $Ci(\mathcal{M}) = Ci_{ac}(\mathcal{M})$.

Finally, let $\mathcal{M} \cup p$ be an extension of $\mathcal{M}$ with p in general position. We define the *extension vector* $Ext_p \in \mathbb{R}^{\Delta(\mathcal{M})}$ of p by

$$Ext_p := \sum_{\substack{\sigma \in \Delta(\mathcal{M}) \\ p \in conv_{\mathcal{M}\cup p}(\sigma)}} e_\sigma .$$

Observe that different extensions can produce the same extension vector. An extension is interior if and only if its extension vector is non-zero.

We fix the standard inner product $\langle \cdot , \cdot \rangle$ on $\mathbb{R}^{\Delta(\mathcal{M})}$. An r-subset σ of E is a basis of $\mathcal{M}$ if and only if its complementary $E \setminus \sigma$ is a basis in the dual oriented matroid $\mathcal{M}^*$, which has rank $(|E| - r)$. Thus, we can identify $\Delta(\mathcal{M})$ and $\Delta(\mathcal{M}^*)$ by complementarity, and this induces an identification of $\mathbb{R}^{\Delta(\mathcal{M})}$ with $\mathbb{R}^{\Delta(\mathcal{M}^*)}$. From duality between circuits and cocircuits it follows that, under this identification, $Ci(\mathcal{M}) = Co(\mathcal{M}^*)$ and $Ci_{ac}(\mathcal{M}) = Co_{int}(\mathcal{M}^*)$. The utility of the above notation becomes clear in the following statement. Observe that the equations $\langle Co_\tau, v_T \rangle = 0$ of parts (b) and (c) are the same as saying that T is a multi-triangulation, in the sense of Definition 2.7.

PROPOSITION 3.1. *Let T be a collection of r-simplices of an oriented matroid $\mathcal{M}$ of rank r. Let $v_T \in \mathbb{R}^{\Delta(\mathcal{M})}$ be its characteristic vector. Then, the following conditions are equivalent:*

(a) *T is a triangulation of $\mathcal{M}$.*
(b) *$\langle Co_\tau, v_T \rangle = 0$ for every interior cocircuit vector Co_τ and $\langle Ext_p, v_T \rangle = 1$ for some interior extension $\mathcal{M} \cup p$ of $\mathcal{M}$ in general position.*
(c) *$\langle Co_\tau, v_T \rangle = 0$ for every interior cocircuit vector Co_τ and $\langle Ext_p, v_T \rangle = 1$ for every interior extension $\mathcal{M} \cup p$ of $\mathcal{M}$ in general position.*

PROOF. The "extension equation" $\langle Ext_p, v_T \rangle = 1$ means the same as "T covers p exactly once". The "cocircuit equations" $\langle Co_\tau, v_T \rangle = 0$ for interior co-rank 1 simplices are weaker than the oriented pseudo-manifold property. This gives the implication from (a) to (b), using characterization (c) of triangulations in Theorem 2.4. The implication from (b) to (c) follows from Lemma 2.8, since the cocircuit equations are a reformulation of being a multi-triangulation.

Suppose now that T is in the conditions of (c). By Theorem 2.4 we only need to prove that T satisfies the oriented pseudo-manifold property. Let $\tau = \{a_1, \dots, a_{r-1}\}$ be a codimension-one simplex of $\mathcal{M}$ not contained in a facet and let $C = (C^+, C^-)$ be a cocircuit vanishing on τ. From the cocircuit equations in part (c) it follows that the number of simplices of T of the form $\tau \cup a$ with $a \in C^+$ equals the number of those with $a \in C^-$. The oriented pseudo-manifold property will be established if we prove that this number is at most one.

If the number is at least two then let $a, b \in C^+$ such that $\tau \cup a$ and $\tau \cup b$ are in T. The lexicographic extensions by $[a_1^+, \dots, a_{r-1}^+, a^+]$ and $[a_1^+, \dots, a_{r-1}^+, b^+]$ coincide: By definition, their cocircuit signatures could only differ on the cocircuits vanishing on τ, but at these cocircuits the sign of a and b is the same. Hence, this interior and in general position lexicographic extension is contained in two simplices of T, violating (c). □

PROPOSITION 3.2. *Let $\mathcal{M}$ be a rank r oriented matroid on E. Let $\mathcal{M}^*$ be the dual oriented matroid. Let $\mathcal{M} \cup p$ an extension of $\mathcal{M}$ in general position. Then,*

(i) *For any extension* $\mathcal{M}^* \cup p^*$ *of* $\mathcal{M}^*$ *in general position, there is at most one* r*-simplex* σ *of* $\mathcal{M}$ *such that* $p \in conv_{\mathcal{M}}(\sigma)$ *and* $p^* \in conv_{\mathcal{M}^*}(E \setminus \sigma)$*. That is,* $\langle Ext_p, Ext_{p^*} \rangle \leq 1$*, under the identification of* $\Delta(\mathcal{M})$ *and* $\Delta(\mathcal{M}^*)$*.*

(ii) *If* p *is interior, then there exists a* p^* *for which equality holds in the previous equation.*

(iii) $\langle Ci_\rho, Ext_p \rangle = 0$*, for every acyclic circuit vector* Ci_ρ *of* $\mathcal{M}$*.*

PROOF. Suppose that there were two r-simplices $\sigma_1 \neq \sigma_2$ in $\mathcal{M}$ with $p \in conv_{\mathcal{M} \cup p}(\sigma_i)$ and $p^* \in conv_{\mathcal{M}^* \cup p^*}(E \setminus \sigma_i)$, for $i = 1, 2$. By the general position assumption, $(\{p\}, \sigma_1)$ and $(\{p\}, \sigma_2)$ are circuits in $\mathcal{M}$. Hence, there are τ_1 and τ_2 with $\sigma_1 \setminus \sigma_2 \subseteq \tau_1 \subseteq \sigma_1$ and $\sigma_2 \setminus \sigma_1 \subseteq \tau_2 \subseteq \sigma_2$ such that (τ_1, τ_2) is a vector in $\mathcal{M}$. The same argument (in the dual) implies that there are τ_1^* and τ_2^* with $\sigma_2 \setminus \sigma_1 \subseteq \tau_1^* \subseteq E \setminus \sigma_1$ and $\sigma_1 \setminus \sigma_2 \subseteq \tau_2^* \subseteq E \setminus \sigma_2$ and such that (τ_1^*, τ_2^*) is a vector in $\mathcal{M}^*$, i.e. a covector in M. These violates orthogonality of vectors and covectors, which proves (i).

If p is interior then there is an r-simplex σ of $\mathcal{M}$ for which p lies in $conv_{\mathcal{M} \cup p}(\sigma)$. Let $\sigma^* := \{a_1, \dots, a_{|E|-r}\}$ be the complement of σ in E. To prove (ii), take p^* to be the lexicographic extension by the element $p^* := [a_1^+, \dots, a_{|E|-r}^+]$.

For (iii), we can assume that $p \in conv_{\mathcal{M} \cup p}(\rho)$, since otherwise the inner product is clearly zero. If this is the case, the value of the inner product will be the same in $\mathcal{M} \cup p$ and in $\mathcal{M}' := (\mathcal{M} \cup p)(\rho \cup p)$, which is acyclic. In the oriented matroid $\mathcal{M}'$ the equation $\langle Ci_\rho, Ext_p \rangle = 0$ follows from parts (i) and (ii) using the fact that the positive and negative parts of Ci_ρ are the characteristic vectors of the two only triangulations of $\mathcal{M}(\rho)$, shown in Proposition 2.13. □

COROLLARY 3.3. *Let* $\mathcal{M}$ *be an oriented matroid on a set* E *and let* $\mathcal{M}^*$ *be its dual oriented matroid. Let* $\mathcal{M}^* \cup p^*$ *be an interior extension of* $\mathcal{M}^*$ *in general position. Then, the following is a triangulation of* $\mathcal{M}$*:*

$$\{\sigma \in \Delta(\mathcal{M}) : p^* \in conv\mathcal{M}^* \cup p^*(E \setminus \sigma).$$

In other words, the extension vector Ext_{p^*} *in* $\mathbb{R}^{\Delta(\mathcal{M}^*)}$ *is the characteristic vector of a triangulation of* $\mathcal{M}$*, under the identification between* $\Delta(\mathcal{M}^*)$ *and* $\Delta(\mathcal{M})$*.*

PROOF. Straightforward from Propositions 3.1 and 3.2, taking into account that the interior cocircuit vectors of $\mathcal{M}$ correspond to the acyclic circuit vectors of $\mathcal{M}^*$ in the identification of $\Delta(\mathcal{M})$ and $\Delta(\mathcal{M}^*)$. □

DEFINITION 3.4. The triangulations of $\mathcal{M}$ obtained by interior extensions in general position of the dual oriented matroid $\mathcal{M}^*$, as in Corollary 3.3, are called *lifting triangulations*. Those obtained by lexicographic extensions are called *lexicographic triangulations*.

Lifting triangulations and in particular lexicographic ones will play an important role in the rest of the paper. The placing triangulations introduced in Section 2.3 were an example of them and Chapter 5 is explicitly devoted to their study. A geometric interpretation of lifting triangulations is given in Section 4.1. In particular, Remark 4.4 shows how to understand lexicographic triangulations in terms of "pushings" and "pullings", as was done in [**24**].

3.2. The affine span of characteristic vectors of triangulations

The following two results are dual to one another and inspired by Theorem 2.2 in [**14**]; actually our proof of Theorem 3.5 is essentially taken from that paper, except that it has been restructured and there are the obvious changes in notation and dualization. The role of regular triangulations is played here by lexicographic extensions.

THEOREM 3.5. *Let $\mathcal{M}$ be an oriented matroid of rank r on a set E. Let v be any vector in $\mathbb{R}^{\Delta(\mathcal{M})}$. The following properties are equivalent:*

(i) *v is a linear combination of the acyclic circuit vectors Ci_ρ of $\mathcal{M}$.*
(ii) *$\langle v_{T^*}, v\rangle = 0$ for the characteristic vector v_{T^*} of every triangulation of the dual oriented matroid $\mathcal{M}^*$.*
(iii) *$\langle Ext_p, v\rangle = 0$ for every extension in general position $\mathcal{M} \cup p$ of $\mathcal{M}$.*
(iv) *$\langle Ext_p, v\rangle = 0$ for every positive lexicographic extension in general position. I.e. for every extension of the form $p := [b_1^+, \ldots, b_r^+]$ where $\{b_1, \ldots, b_r\}$ is an r-simplex of $\mathcal{M}$.*

Moreover, if v is integer the combination in part (i) can be taken with integer coefficients.

PROOF. If v is a linear combination of acyclic circuit vectors of $\mathcal{M}$ then it is a linear combination of interior cocircuit vectors of $\mathcal{M}^*$. Thus, (ii) follows from (i) by Proposition 3.1. Also, (iii) is the restriction of (ii) to the case of T^* being a lifting triangulation, and (iv) is a restriction of (iii). Thus, we only need to prove the implication from (iv) to (i).

Let v be in the conditions of (iv). Let c_σ denote the coefficient of v in the coordinate of an r-simplex σ. We will use double induction on $n = |E|$ and $r = rank(\mathcal{M})$. In particular, we assume the statement to be true for the deletion $\mathcal{M} \setminus a$ and the contraction $\mathcal{M}/a$, where a is any element of E. We suppose that a is not a loop, since otherwise the inductive step is trivial. We do the proof in three steps:

Step 1: if $c_\sigma = 0$ for every simplex σ containing a, then v can be considered a vector in $\mathbb{R}^{\Delta(\mathcal{M}\setminus a)}$. We suppose that a is not a coloop in $\mathcal{M}$, since otherwise $v = 0$.

The vector v satisfies (iv) in $\mathcal{M}\setminus a$ because every positive lexicographic extension in general position of $\mathcal{M} \setminus a$ can be extended to a positive lexicographic extension in general position of $\mathcal{M}$, by Lemma 1.9 (note that $E \setminus \{a\}$ spans $\mathcal{M}$, since a is not a coloop). By inductive hypothesis, v is a linear combination of acyclic circuit vectors of $\mathcal{M} \setminus a$ (and an integer combination if v is integer). Every acyclic circuit vector of $\mathcal{M} \setminus a$ is also an acyclic circuit vector in $\mathcal{M}$, if a is not a coloop, which means v is in the conditions of (i).

Step 2: if $c_\sigma = 0$ for every simplex σ not containing a and such that $\sigma \cup \{a\}$ is acyclic, then we can write:

$$v = \sum_{\substack{\sigma : a \notin \sigma \\ \sigma\cup\{a\} \text{ is cyclic}}} c_\sigma e_\sigma + \sum_{\sigma : a \in \sigma} c_\sigma e_\sigma .$$

Let us call v_1 and v_2 the two sums in this expression, respectively.

The r-simplices of $\mathcal{M}$ containing a are naturally identified with the $(r-1)$-simplices of $\mathcal{M}/a$. Thus, v_2 can be regarded as a vector in $\mathbb{R}^{\Delta(\mathcal{M}/a)}$, removing a from each simplex. We claim that v_2 is in the conditions of (iv), in the contracted oriented matroid $\mathcal{M}/a$. Indeed, for any lexicographic extension by $p' := [b_1^+, \ldots, b_{r-1}^+]$ of $\mathcal{M}/a$, we can consider the lexicographic extension by $p := [a^+, b_1^+, \ldots, b_{r-1}^+]$ of $\mathcal{M}$. All the simplices σ with non-zero entry in the extension vector Ext_p satisfy that $\sigma \cup \{a\}$ is acyclic. Thus, $\langle Ext_p, v_1\rangle = 0$. Since also

$\langle Ext_p, v\rangle = 0$ we conclude that $\langle Ext_p, v_2\rangle = 0$. But the simplices of Ext_p containing a are the same as the simplices of $Ext_{p'}$, and thus $\langle Ext_{p'}, v_2\rangle = 0$.

By inductive hypothesis, v_2 is a linear combination of the acyclic circuit vectors of $\mathcal{M}/a$, and an integral combination if v_2 is integral. Given any acyclic spanning r-set ρ in $\mathcal{M}/a$, $\rho \cup \{a\}$ is a spanning acyclic $(r+1)$-set in $\mathcal{M}$. Moreover, the acyclic circuit vector $Ci_{\rho\cup\{a\}}$ in $\mathcal{M}$ is obtained from the acyclic circuit vector Ci_ρ of $\mathcal{M}/a$ by just considering each simplex of $\mathcal{M}/a$ as a simplex of $\mathcal{M}$ containing a and adding an entry $+1$ or -1 to the coordinate of ρ, in case that ρ is independent in $\mathcal{M}$. In other words, the expression of v_2 as a linear combination of acyclic circuit vectors of $\mathcal{M}/a$ translates into an expression of v_2 as a linear combination L of acyclic circuit vectors of $\mathcal{M}$ plus an integer combination of the form $\sum_{\sigma:a\notin\sigma} c'_\sigma e_\sigma$. Subtracting L from v, we get another vector

$$v' \quad = \sum_{\sigma:a\notin\sigma} c'_\sigma e_\sigma$$

in the conditions of (iv). Both $v - v'$ (by construction) and v' (by step 1) are linear combinations of acyclic circuit vectors, and integer combinations if v is integral.

Step 3: Suppose now we have an arbitrary v. For each r-simplex σ for which the spanning $(r+1)$-set $\sigma \cup \{a\}$ is acyclic, we consider the acyclic circuit vector $Ci_{\sigma\cup\{a\}}$ with sign given so that the coefficient of σ equals 1. All the other nonzero coefficients in $Ci_{\sigma\cup\{a\}}$ correspond to simplices containing σ. Hence, subtracting from v the vector $c_\sigma Ci_{\sigma\cup\{a\}}$ for every such σ we get another vector v' in which the coefficients corresponding to those simplices are zero. The difference $v-v'$ is a linear combination of acyclic circuit vectors, and an integer combination if v is integer. This implies that v' is still in the conditions of (iv), by part (iii) of Proposition 3.2, and it is integer if v is integer. By Step 2, v' is a linear combination of acyclic circuit vectors, and an integer combination if v is integer. □

The lifting triangulations obtained with positive lexicographic extensions are the *placing* triangulations (see Proposition 2.11). With this, the previous theorem dualizes to:

COROLLARY 3.6. *Let $\mathcal{M}$ be an oriented matroid of rank r on a set E. Let h be any vector in $\mathbb{R}^{\Delta(\mathcal{M})}$. The following properties are equivalent:*

(i) *h is a linear combination of the interior cocircuit vectors Co_τ of $\mathcal{M}$ and if h is integer the combination has integer coefficients.*
(ii) *$\langle h, v_T\rangle = 0$ for every triangulation T of $\mathcal{M}$.*
(iii) *$\langle h, v_T\rangle = 0$ for every placing triangulation T of $\mathcal{M}$.* □

COROLLARY 3.7. *Let $\mathcal{M}$ be an oriented matroid. Then,*

(i) *The characteristic vector of any triangulation is an affine combination of characteristic vectors of placing triangulations.*
(ii) *The affine span of all the characteristic vectors of triangulations of $\mathcal{M}$ is defined by the interior cocircuit equations $\langle Co_\tau, \cdot\ \rangle = 0$ and any non-homogeneous affine equation satisfied on every characteristic vector (e.g., the equation $\langle Ext_p, \cdot\ \rangle = 1$, for any interior extension $\mathcal{M} \cup p$ of $\mathcal{M}$ in general position).*
(iii) *The difference $v_T - v_{T'}$ of the characteristic vectors of two triangulations of $\mathcal{M}$ is an integer combination of acyclic circuit vectors of $\mathcal{M}$.*

PROOF. Part (ii) follows from the equivalence of parts (i) and (iii) of Corollary 3.6: an affine subspace not containing the origin can be obtained intersecting its linear span with any affine hyperplane containing it and not containing the origin.

The equivalence of parts (i) and (ii) in Corollary 3.6 implies that the affine spans of characteristic vectors of placing triangulations and of all characteristic vectors of triangulations coincide. This proves part (i).

For part (iii), first Proposition 3.1 implies $\langle Ext_p, v_T - v_{T'}\rangle = 0$ for any extension in general position of $\mathcal{M}^*$. Then, Theorem 3.5 implies that $v_T - v_{T'}$ is an integer combination of acyclic circuit vectors. □

This leads to a stronger version of Proposition 3.1, which in particular gives characterization (g) of triangulations in Theorem 2.4.

THEOREM 3.8. *Let T be a collection of r-simplices of an oriented matroid $\mathcal{M}$ of rank r. Let $v_T \in \mathbb{R}^{\Delta(\mathcal{M})}$ be its characteristic vector. Then, the following conditions are equivalent:*

(a) *T is a triangulation of $\mathcal{M}$.*
(b) *$\langle Co_\tau, v_T\rangle = 0$ for every interior cocircuit vector Co_τ and $\langle v_{T^*}, v_T\rangle = 1$ for some triangulation v_{T^*} of the dual $\mathcal{M}^*$.*
(c) *$\langle Co_\tau, v_T\rangle = 0$ for every interior cocircuit vector Co_τ and $\langle v_{T^*}, v_T\rangle = 1$ for every triangulation v_{T^*} of the dual $\mathcal{M}^*$.*
(d) *T satisfies the oriented pseudo-manifold property and for every triangulation T^* of the dual oriented matroid $\mathcal{M}^*$ there is a unique simplex in T whose complement is in T^*.*

PROOF. Statement (c) implies part (b) of Proposition 3.1, taking any lifting triangulation T^*). With the same trick, part (c) of Proposition 3.1 implies statement (b). This proves (c)⇒(a)⇒(b).

Let T^* be a triangulation of $\mathcal{M}^*$. By (the dual of) Proposition 3.1, $\langle v_{T^*}, v_T\rangle = 1$ for every lifting triangulation v_T of $\mathcal{M}$. By part (i) of Corollary 3.7 the same equation holds for every triangulation of $\mathcal{M}$, lifting or not. This means that the affine equation $\langle v_{T^*}, \cdot\rangle = 1$ is a non-homogeneous affine equation satisfied on every triangulation of $\mathcal{M}$. With this, (b)⇒(c) follows from part (ii) of Corollary 3.7.

That (d) implies (c) is trivial. That (c) and (a) imply (d) is trivial too, since all triangulations have the oriented pseudo-manifold property. □

Part (ii) of Corollary 3.7 implies the relation $D = N - R - 1$, between the number N of r-simplices (bases) of $\mathcal{M}$, the rank R of the linear span of its interior cocircuit vectors and the dimension D of the affine span of all the characteristic vectors of triangulations of $\mathcal{M}$. If the oriented matroid $\mathcal{M}$ is uniform we clearly have $N = \binom{n}{r}$; we also can give explicit formulas for the other quantities, as is done in [**14**].

LEMMA 3.9. *Let $\mathcal{M}$ be a uniform oriented matroid. Then, every interior cocircuit vector of $\mathcal{M}$ is the difference of two interior extension vectors and every non-interior cocircuit vector is (up to sign reversal) an interior extension vector.*

PROOF. Let C be a cocircuit. Let $\tau = \{a_1, \dots, a_{r-1}\}$ be the unique $(r-1)$-simplex in which the cocircuit vanishes. Let a_r be an element not in τ. Consider the two lexicographic extensions $p^\epsilon := [a_1^+, \dots, a_{r-1}^+, a_r^\epsilon]$, for $\epsilon \in \{+,-\}$. Since their cocircuit signatures only differ at C and $-C$, any r-simplex not containing τ either contains both p^+ and p^- or none of them in its convex hull. On the other hand, for any $a \notin \tau$ and for any of the two values of ϵ, $p^\epsilon \in \mathrm{conv}_{\mathcal{M} \cup p^\epsilon}(\tau \cup \{a\})$ if and only if $C(a) = \epsilon$. Hence $Ext_{p^+} - Ext_{p^-} = Co_\tau$. If τ is interior then the two extensions are interior, while otherwise only one of the two extensions is interior and the extension vector of the other one is zero. □

THEOREM 3.10. *Let $\mathcal{M}$ be an acyclic uniform oriented matroid of rank r on n elements. Let a_1 be any element. Let $H_\mathcal{M}$ be the affine span of the characteristic vectors of triangulations of $\mathcal{M}$. Then,*

(i) *The cocircuit equations $\langle Co_\tau, \cdot \rangle = 0$ if τ is interior and $\langle Co_\tau, \cdot \rangle = 1$ otherwise, for the $(r-1)$-subsets τ with $a_1 \notin \tau$ form a basis for the space of affine equations valid on $H_\mathcal{M}$. Thus, $R + 1 = \binom{n-1}{r-1}$.*

(ii) *The (always acyclic) circuit vectors Ci_ρ, for the $(r+1)$-subsets ρ with $a_1 \in \rho$ form a basis for the linear space parallel to $H_\mathcal{M}$. Thus, $D = \binom{n-1}{r}$.*

(iii) *The linear spans $lin(Co(\mathcal{M}))$ and $lin(Ci(\mathcal{M}))$ of the cocircuit and circuit vectors of $\mathcal{M}$ are orthogonal complements in $\mathbb{R}^{\Delta(\mathcal{M})}$, of dimensions $\binom{n-1}{r-1}$ and $\binom{n-1}{r}$.*

PROOF. In the realizable case, parts (i) and (ii) are Theorem 2.4 in [**14**] and part (iii) is Theorem 5.1 in that paper. The same proofs work here. □

Part (iii) of Theorem 3.10 holds also if $\mathcal{M}$ is uniform but not acyclic: Then $\mathcal{M}$ is totally cyclic and $\mathcal{M}^*$ is uniform and acyclic, so we have part (iii) for $\mathcal{M}^*$ and, by duality, for $\mathcal{M}$. But in this case parts (i) and (ii) change: one will have $R = \binom{n-1}{r-1}$ (because now all the cocircuit equations are interior, so one extra non-homogeneous equation is needed to define $H_\mathcal{M}$) and $D + 1 = \binom{n-1}{r}$.

Theorem 3.10 implies that the interior cocircuit equations (the first ones in part (c) of Proposition 3.1) follow from the extension equations (the second ones) in the uniform case. This is Corollary 5.5 in [**14**] in the realizable case. If $\mathcal{M}$ is not uniform then this is not true and $lin(Co(\mathcal{M}))$ and $lin(Ci(\mathcal{M}))$ are not orthogonal complements. One can only prove that any of them contains the orthogonal complement of the other [**14**, Proposition 5.3].

3.3. Mutations versus geometric bistellar flips

Both for extensions of an oriented matroid and for triangulations of a point configuration there are notions of a "local" or "elementary" change between two of them. These are, respectively, the so-called *mutations* and *geometric bistellar flips*. It is not surprising that these two concepts be related to one another under the duality of triangulations and extensions depicted in the previous sections. Here we explore this duality. We take the following as a definition:

DEFINITION 3.11. Let $\mathcal{M}$ be an oriented matroid. Let $\mathcal{M} \cup p_1$ and $\mathcal{M} \cup p_2$ be two extensions of $\mathcal{M}$ in general position and let T_1 and T_2 two triangulations of $\mathcal{M}$.

(i) We say that p_1 and p_2 differ by a *mutation* if their cocircuit signatures differ only in one pair of opposite cocircuits. We say that the mutation is *supported* on those cocircuits.

(ii) We say that T_1 and T_2 differ by a *geometric bistellar flip* (or a flip, for short) if the difference of their characteristic vectors is a sum of acyclic circuit vectors which are supported on the same circuit C. (We say that a circuit vector Ci_ρ is *supported* on a circuit $C = (C^+, C^-)$ if C is the circuit contained in ρ, oriented so that the coefficient of $\rho\setminus a$ in Ci_ρ equals the sign of $C(a)$ for every $a \in \underline{C}$). We say that the flip is *supported* on C.

What we call mutation was introduced by Fukuda and Tamura (1988) with the name of *flipping* (see [**11**, Section 7.3]). Only if $\mathcal{M}$ is uniform it is usually called a mutation [**11**, Definition 7.3.8]. We prefer not to use the name flipping in order to not create confusion with the flips in triangulations. Mutations correspond to moving the general position element p_1 to an "almost-general" position in which the circuits containing it have at least r elements and then perturbing it back to general position:

PROPOSITION 3.12. *Let $\mathcal{M}$ be an oriented matroid. Let $C = (C^+, C^-)$ be a cocircuit of $\mathcal{M}$. Let C_0 denote the complement of the support of C and let $\mathcal{M}_0$ denote the restriction of $\mathcal{M}$ to C_0. Let $\mathcal{M} \cup p$ be an extension of $\mathcal{M}$ whose cocircuit signature is zero only on the cocircuit C (and its opposite). Let $a \in \underline{C}$.*

(i) *The only extensions of $\mathcal{M}$ which are perturbations of $\mathcal{M} \cup p$ are the two lexicographic perturbations of $\mathcal{M} \cup p$ obtained as $p_{a^+} := [p^+, a^+]$ and $p_{a^-} := [p^+, a^-]$.*

(ii) *Let $\{\tau_1, \dots, \tau_l\}$ be the bases of $\mathcal{M}_0$ (i.e., the $(r-1)$-simplices contained in C_0) satisfying $p \in conv_{\mathcal{M}\cup p}(\tau_i)$. Let $a \in C^+$. Consider the cocircuit vectors $Co_{\tau_1}, \dots, Co_{\tau_l}$ with signs given so that the coefficient of the simplex $\tau_i \cup \{a\}$ is positive. Then,*

$$\sum_{i=1}^{l} Co_{\tau_i} = Ext_{p_{a^+}} - Ext_{p_{a^-}}.$$

(iii) *$\mathcal{M} \cup p_{a^+}$ and $\mathcal{M} \cup p_{a^-}$ differ by a mutation supported on C. Moreover, every pair of extensions which differ by a mutation can be obtained in this way.*

PROOF. The proof of (i) is straightforward: clearly, p_{a^+} and p_{a^-} are well-defined extensions which are perturbations of p. In the other hand, any perturbation of p is determined by its value on the cocircuit C.

For proving (ii), let σ be an arbitrary r-simplex of $\mathcal{M}$ and let us see that its coefficient in the right hand side equals the one in the left hand side. Suppose that the left hand side is non-zero. This implies that σ contains one of the simplices τ_i; because of part (i) we can assume that $\sigma = \tau_i \cup a$ without loss of generality, and then it becomes clear that the coefficient of σ in both the right hand side and the left hand side is 1.

Reciprocally, suppose that the coefficient of σ in the right hand side is non-zero. This means that exactly one of the two extension elements p_{a^+} and p_{a^-} is contained in the convex hull of σ and, in particular, that $p \in conv_{\mathcal{M}\cup p}(\sigma)$, because of part (ii) of Lemma 1.8. Also, part (v) of Lemma 1.1 tells us that the signatures of the two extensions must differ in a cocircuit vanishing in a facet of σ, that is, that σ contains an $(r-1)$-simplex contained in C^0. This two facts together imply that σ contains one of the simplices τ_i. But then, we can assume without loss of generality that $\sigma = \tau_i \cup a$ and, as before, conclude that the coefficient of σ in both sides of the equation equals one.

For (iii), it is clear that $\mathcal{M} \cup p_{a^+}$ and $\mathcal{M} \cup p_{a^-}$ differ by a mutation supported on C. Reciprocally, Lemma 7.3.3 of [**11**] says that whenever we have two extensions

$\mathcal{M} \cup p^+$ and $\mathcal{M} \cup p^-$ in general position whose cocircuit signatures agree in every cocircuit except C and its opposite, there is a third extension $\mathcal{M} \cup p$ whose cocircuit signature coincides with the first two except in C and its opposite, where it has the third possible value (zero in our case). This is the $\mathcal{M} \cup p$ in our statement and $\mathcal{M} \cup p^+$ and $\mathcal{M} \cup p^-$ are perturbations of it. □

We now look at geometric bistellar flips. Our Definition 3.11 of them is rather abstract, while for the case of triangulations of a point configuration a more geometric definition exists. This is for example what Gel'fand et al. [**19**, pages 231–233] call a *modification* of a triangulation, and is a concept which appears quite often in recent literature on triangulations of polytopes (see [**13, 15, 23, 25, 28, 32, 36**]).

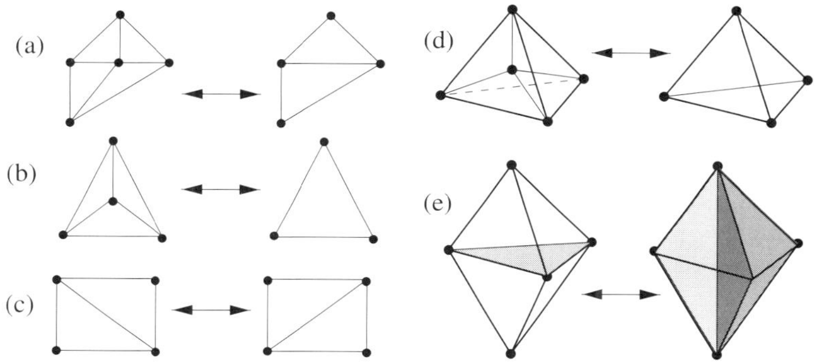

FIGURE 3.1. Some examples of geometric bistellar flips.

Part (i) of the following statement express the geometric definition of a flip in oriented matroid terms, and part (ii) says that this more geometric definition is equivalent to our abstract one. Examples of flips appear in Figure 3.1. Parts (a), (b) and (c) of the Figure show the three possible types of flips in dimension 2 (rank 3). The one in part (a) is degenerate in the sense that it is supported in a non-full-rank circuit. Parts (d) and (e) show the two possible non-degenerate flips in dimension 3 (rank 4). Remember that we use the notation $A \cdot B$ for the *join* of two collections of simplices, defined as

$$A \cdot B := \{\sigma \cup \tau : \sigma \in A, \tau \in B\}.$$

Also, we recall from Proposition 2.13 that the restriction of an oriented matroid to the support of an acyclic circuit $C = (C^+, C^-)$ has only the following two triangulations:

$$T_C^+ := \{\underline{C} \setminus \{e\} : e \in C^+\} \qquad \text{and} \qquad T_C^- := \{\underline{C} \setminus \{e\} : e \in C^-\}.$$

PROPOSITION 3.13. *Let $\mathcal{M}$ be an oriented matroid and let $C := (C^+, C^-)$ be an acyclic circuit (that is, a circuit with non-empty positive and negative parts) of $\mathcal{M}$. Let T be a triangulation of $\mathcal{M}$. Then:*

(i) *Suppose that the triangulation T_C^- of $\underline{C}$ is a subcomplex of T (that is, every simplex of T_C^- is contained in a simplex of T) and that the links in*

T of all the simplices of T_C^- coincide. In these conditions, let L be the link in T of the simplices of T_C^-. Then, T contains $T_C^- \cdot L$ and

$$T' := T \setminus (T_C^- \cdot L) \cup (T_C^+ \cdot L)$$

is a triangulation of $\mathcal{A}$.

(ii) *T and T' differ by a geometric bistellar flip supported on the circuit C. Moreover, every pair of triangulations which differ by a geometric bistellar flip arise in this way.*

PROOF. (i) The fact that T contains $T_C^- \cdot L$ is clear, since L is the link of every simplex of T_C^-. All the maximal simplices in T_C^- and in T_C^+ span the same flat in $\mathcal{M}$, namely the flat spanned by $\underline{C}$. Hence, $T_C^+ \cdot L$ and T' are collections of r-simplices of $\mathcal{M}$, since T and $T_C^- \cdot L$ are.

We have to prove that T' is a triangulation. Let v and v' be the characteristic vectors of T and T' respectively. Let v^+ and v^- be those of $T_C^+ \cdot L$ and $T_C^- \cdot L$. Let $\sigma_1, \dots, \sigma_l$ be the maximal simplices in L. Then, we have

$$v - v' = v^- - v^+ = \sum_{i=1}^{l} C i_{\sigma_i \cup \underline{C}}.$$

By part (iii) of Proposition 3.2, $\langle v, Ext_p\rangle = \langle v', Ext_p\rangle$ for any interior extension in general position. We now prove that $\langle Co_\tau, v\rangle = \langle Co_\tau, v'\rangle$ for every interior cocircuit vector Co_τ. These equalities, together with the previous ones, imply that v' is in the conditions of part (c) of Proposition 3.1.

Let τ be an interior $(r-1)$-simplex and let Co_τ be one of the to opposite interior cocircuit vectors defined by τ. Since $v - v' = v^- - v^+$ we only need to prove that $\langle Co_\tau, v^+\rangle = \langle Co_\tau, v^-\rangle$. If τ is not contained in any simplex of $T_C^+ \cdot L$ or $T_C^- \cdot L$ then $\langle Co_\tau, v^+\rangle = \langle Co_\tau, v^-\rangle = 0$ trivially. Otherwise our proof will rely only on the facts that T_C^+ and T_C^- are triangulations of $\mathcal{M}(\underline{C})$ and that L is a triangulation of $\mathcal{M}/\underline{C}$. In particular, the case of τ contained in a simplex of $T_C^+ \cdot L$ or of $T_C^- \cdot L$ will be analogous, and we deal only with the first one.

By a rank argument, τ contains a maximal simplex of either T_C^+ or L. In the first case, let σ be such a simplex, so that $\sigma = \underline{C} \setminus a$ for some $a \in C^+$. Clearly $\langle Co_\tau, v^-\rangle = 0$ because every maximal simplex of $T_C^- \cdot L$ contains C^- and τ contains C^+. On the other hand, $\langle Co_\tau, v^+\rangle = 0$ since L is a triangulation of $\mathcal{M}/\sigma$.

In the second case, let σ be the maximal simplex of L contained in τ. C is still a circuit in $\mathcal{M}/\sigma$, since σ joined to a proper subset of $\underline{C}$ is independent. By induction on the rank, in the oriented matroid $\mathcal{M}/\sigma$ we have that $\mathrm{link}_T(\sigma)$ and $\mathrm{link}_{T'}(\sigma)$ are triangulations differing by a flip on C. In particular, the difference of their characteristic vectors is orthogonal to the interior cocircuit vector associated to $\tau \setminus \sigma$. Hence, $\langle Co_\tau, v^+ - v^-\rangle = 0$, as desired.

(ii) For each simplex $\tau \in L$, $\underline{C} \cup L$ is a spanning $(r+1)$-subset whose circuit vector is precisely the difference of the incidence vectors of $T_C^+ \cdot \tau$ and $T_C^- \cdot \tau$. This proves that $v'_T - v_T$ is a sum of acyclic circuit vectors supported on the circuit C.

For the converse, let T and T' be two triangulations which differ by a flip supported on the acyclic circuit $C = (C^+, C^-)$. Let r and k be the ranks of $\mathcal{M}$ and C. Each circuit vector supported on C is the difference of the incidence vectors of $T_C^+ \cdot \tau$ and $T_C^- \cdot \tau$, for some $(r-k)$-simplex τ. Thus, the fact that T and T' are triangulations and their incidence vectors differ by a sum of circuit vectors supported on C implies that T_C^+ is a subcomplex of one of them and T_C^-

a subcomplex of the other. Suppose that T_C^- is a subcomplex of T and T_C^+ of T'. Characterization (f) of Theorem 2.4 implies that T does not contain any simplex of T_C^+ as a face and T' does not contain one of T_C^-. Thus, the fact that we pass from T to T' by a sum of acyclic vectors supported on C implies that all the simplices of T_C^- have the same link L in T, which becomes the link of all the simplices of T_C^+ in T'. This finishes the proof. □

The following statement gives the relation between flips of lifting triangulations and mutations of the associated extensions of the dual:

THEOREM 3.14. *Let $\mathcal{M}$ be an oriented matroid of rank r in n elements, with dual $\mathcal{M}^*$. Let $\mathcal{M} \cup p_1$ and $\mathcal{M} \cup p_2$ be two interior extensions of $\mathcal{M}$ in general position which differ by a mutation and let T_1 and T_2 be the corresponding lifting triangulations of $\mathcal{M}^*$. Then, either $T_1 = T_2$ or T_1 and T_2 differ by a geometric bistellar flip.*

PROOF. Consider the extension $\mathcal{M} \cup p$ of $\mathcal{M}$ of which p_1 and p_2 are perturbations (as in Proposition 3.12). Part (ii) of Proposition 3.12 says that the extension vectors Ext_{p_1} and Ext_{p_2} differ by a (perhaps empty) sum of cocircuit vectors of $\mathcal{M}$ supported on the same cocircuit; since p_1 and p_2 are both interior, the $(r-1)$-simplices τ_i are not on facets of $\mathcal{M}$, and thus each Co_{τ_i} is an interior cocircuit vector. Duality implies that v_{T_1} and v_{T_2} differ by a (perhaps empty) sum of acyclic circuit vectors supported on the same circuit. Thus, either $T_1 = T_2$ or they differ by a flip. □

Two (related) open questions concerning mutations and flips are whether any two triangulations (resp. any two extensions in general position) of a uniform realizable oriented matroid $\mathcal{M}$ are connected by a sequence of flips (resp. mutations). We will study these questions in more detail in Section 4.3, and see that the answer is negative if $\mathcal{M}$ is not assumed to be realizable or, in the case of triangulations, uniform.

Here we want to mention a sort of algebraic positive answer which was shown for the realizable case in [**6**] and [**14**]. If $\mathcal{M}$ is a uniform oriented matroid, then the link L appearing in Proposition 3.13 is empty and the difference between the characteristic vectors of the two triangulations T and T' which differ by a flip supported on a circuit C equals $Ci_{\underline{C}}$ (with the appropriate sign). Then, part (iii) of Corollary 3.7 says that the difference of the characteristic vectors of any two triangulations of $\mathcal{M}$ is a sum of differences of pairs of characteristic vectors of triangulations differing by flips.

In the non-uniform case this result is not clear: each difference vector between triangulations differing by a flip is a sum of perhaps more than one acyclic circuit vector (as is suggested by the equation in part (ii) of Proposition 3.12).

CHAPTER 4

Subdivisions of Lawrence Polytopes

4.1. Lifting subdivisions. Subdivisions

We have defined lifting triangulations of an oriented matroid $\mathcal{M}$ by means of extensions of the dual oriented matroid $\mathcal{M}^*$. In [**11**, pag. 410] lifting triangulations are defined in a more geometric (but equivalent) way that explains the name "lifting". The idea comes from a paper by Billera and Munson [**9**] although there only lexicographic lifts are considered. We introduce now that definition and show its equivalence with Definition 3.4. The following definition of a *lift* of an oriented matroid is also taken from [**9**] and the same concept appears in [**11**] under the name *one-element lifting*. It is the dual concept to a one-element extension.

DEFINITION 4.1. Let $\mathcal{M}$ be an oriented matroid of rank r on a set E. A *lift* of $\mathcal{M}$ is an oriented matroid $\widehat{\mathcal{M}}$ of rank $r+1$ on a set $E \cup \hat{p}$ such that $\widehat{\mathcal{M}}/\hat{p} = \mathcal{M}$. We say that the lift is *non-cyclic* if $\hat{p}$ does not belong to any positive circuit of $\widehat{\mathcal{M}}$. Equivalently, if $\hat{p}$ belongs to some positive cocircuit of $\widehat{\mathcal{M}}$, by [**11**, Corollary 3.4.6]. We say it is *acyclic* if $\widehat{\mathcal{M}}$ is acyclic.

Given a non-cyclic lift $\widehat{\mathcal{M}}$ of an oriented matroid $\mathcal{M}$, the *lifting polytopal subdivision* (or *lifting subdivision* for short) S of $\mathcal{M}$ associated to the lift is the following collection of subsets of E, to be called *cells* of the subdivision:

$$S := \{A \subseteq E \quad : \quad A \text{ is a facet of } \widehat{\mathcal{M}}\}$$

We say that the subdivision is *simplicial* if all the cells are simplices.

In this definition, the condition on the lift being non-cyclic is exactly what we need to guarantee that S is not empty. The first (trivial) example of a lifting subdivision of $\mathcal{M}$ is given by the lift of $\mathcal{M}^*$ by a coloop, which produces the trivial subdivision $\{E\}$ of $\mathcal{M}$.

PROPOSITION 4.2. *Simplicial lifting subdivisions and lifting triangulations are the same thing. More precisely:*

(i) *Let T be a simplicial lifting subdivision of $\mathcal{M}$ associated with the non-cyclic lift $\widehat{\mathcal{M}}$. The dual $(\widehat{\mathcal{M}})^*$ is an extension of $\mathcal{M}^*$ that we denote by $\mathcal{M}^* \cup p$. Let $\mathcal{M}^* \cup \overline{p}$ be the reorientation of the element p in $\mathcal{M}^* \cup p$. Then, any perturbation of $\mathcal{M}^* \cup \overline{p}$ into general position is an interior extension with associated lifting triangulation of $\mathcal{M}$ equal to T.*

(ii) *Let T be a lifting triangulation of $\mathcal{M}$ associated with an extension $\mathcal{M}^* \cup \overline{p}$ of the dual oriented matroid $\mathcal{M}^*$. Denote by $\mathcal{M}^* \cup p$ the reorientation at the element $\overline{p}$ of $\mathcal{M}^* \cup \overline{p}$. Then, the dual oriented matroid $(\mathcal{M}^* \cup p)^*$ is a lift of $\mathcal{M}$ whose lifting subdivision is T.*

PROOF. Let us first give a different characterization of the cells of a lifting subdivision. A subset $A \subseteq E$ is a cell in the lifting subdivision of the lift $\widehat{\mathcal{M}}$ if and only if $((E \setminus A) \cup \widehat{p}, \emptyset)$ is a cocircuit in $\widehat{\mathcal{M}}$. This in particular implies that $A \cup \widehat{p}$ is a spanning subset of $\widehat{\mathcal{M}}$; thus, A is a spanning subset of $\mathcal{M}$ and $E \setminus A$ is an independent subset of $\mathcal{M}^*$.

The dual of a non-cyclic lift $\widehat{\mathcal{M}}$ of $\mathcal{M}$ is an extension $\mathcal{M}^* \cup p$ of the dual $\mathcal{M}^*$ with p not lying in any positive cocircuit, i.e. lying in some positive circuit. Thus, the reorientation $\mathcal{M}^* \cup \overline{p}$ is an interior extension of $\mathcal{M}^*$. The reciprocal is also true; that is, there is a 1-to-1 correspondence between non-cyclic lifts of $\mathcal{M}$ and interior extensions of $\mathcal{M}^*$, by reorientation of the dual. Under this correspondence, $((E \setminus A) \cup \widehat{p}, \emptyset)$ is a cocircuit of the lift $\widehat{\mathcal{M}}$ if and only if $(E \setminus A, \{\overline{p}\})$ is a circuit of the extension $\mathcal{M}^* \cup \overline{p}$. This implies (ii), since the lifting triangulation T contains by definition precisely those independent sets A of $\mathcal{M}^* \cup \overline{p}$ for which $(A, \{\overline{p}\})$ is a circuit of $\mathcal{M}^* \cup \overline{p}$.

For proving (i) we have the extra difficulty that $\mathcal{M}^* \cup \overline{p}$ may not be an extension in general position. However, we have the following property, which follows from the fact that T is simplicial: any subset $A \in E$ with $\overline{p} \in conv_{\mathcal{M}^* \cup \overline{p}}(A)$ is spanning; in particular we have that $\overline{p} \in relint_{\mathcal{M}^* \cup \overline{p}}(A)$ and thus that $p' \in relint_{\mathcal{M}^* \cup p'}(A)$ for any perturbation $\mathcal{M}^* \cup p'$ of $\mathcal{M}^* \cup \overline{p}$. That is, all the simplices of T are in the lifting triangulation corresponding to p'. The reciprocal follows with the same kind of arguments. □

EXAMPLE 4.3. (Lifting triangulations via lifts)

An example of the equivalence between the two definitions of lifting triangulations is shown in Figure 4.1. Parts (a) and (b) show an oriented matroid $\mathcal{M}$ and its dual $\mathcal{M}^*$, both of rank 2 and realized as vector configurations. Parts (c) and (e) show two different acyclic lifts of $\mathcal{M}$, which have rank 3, realized as point configurations in the plane. Recall that if an oriented matroid is realized by a point configuration $\mathcal{A}$ and p is a point of $\mathcal{A}$ (i.e., an element of $\mathcal{M}$), the contraction $\mathcal{M}/p$ is realized by the vector configuration $\{a - p : a \in \mathcal{A} \setminus \{p\}\}$.

The segments drawn in parts (c) and (e) of the figure are the facets of the lift which do not contain p; that is, the cells of the induced lifting triangulations of $\mathcal{M}$. Parts (d) and (f) of the figure show two extensions of $\mathcal{M}^*$ (noted p) and their opposites (noted $\overline{p}$). It is easily checked that the simplices of part (d) (resp. part(f)) which contain the extension $\overline{p}$ in their convex hulls are the complements of the simplices in the lifting subdivision in part (c) (resp. part (e)).

REMARK 4.4. (Lexicographic triangulations by "pushings and pullings")

In the realizable acyclic case, lexicographic triangulations were characterized by Carl Lee [**24**] as the ones that can be obtained from the trivial subdivision of a polytope by a sequence of *pushings* and *pullings* of points. This description is generalized to oriented matroid triangulations in [**11**, p. 410], still in the acyclic case, but the description works the same in the non-acyclic case. Figure 4.2 shows an example of the process.

Part (a) of the figure shows a certain triangulation of a planar point configuration and part (b) shows how to obtain it by a "lexicographic lift" with the expression $[1^-, 2^-, 3^+]$. Recall that this corresponds to the triangulation associated to the opposite lexicographic extension of the dual, that is, to $[1^+, 2^+, 3^-]$. The lexicographic lift is constructed by adding a coloop p to $\mathcal{M}$ (the apex of the

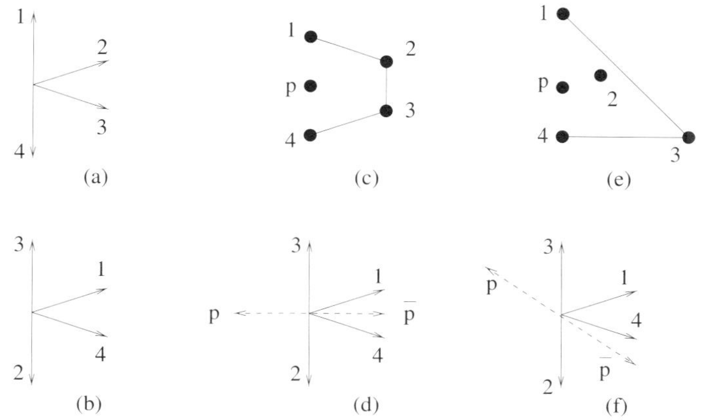

FIGURE 4.1. Lifting triangulations defined by lifts and by extensions of the dual.

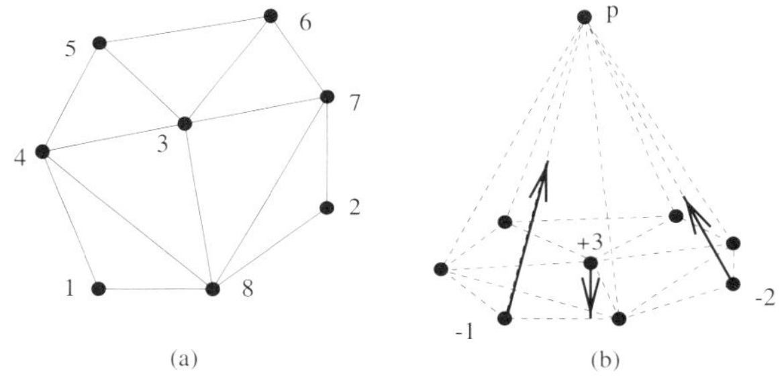

FIGURE 4.2. A lexicographic triangulation and the associated lift.

pyramid in part (b) of the figure) and then perturbing the elements in the order they appear in the lexicographic expression, *pulling* them towards the apex if they have negative sign and *pushing* them away from the apex if they have positive sign.

The definition of *lifting subdivisions* suggests the general concept of a *subdivision* of an oriented matroid. This has to agree with the concept of polytopal subdivision of a polytope if the oriented matroid is polytopal, and with the concept of triangulation in the simplicial case. The following definition is taken from [**11**, page 408], except that there $\mathcal{M}$ is assumed to be acyclic.

DEFINITION 4.5. Let $\mathcal{M}$ be an oriented matroid of rank r on a set E. A nonempty collection S of subsets of E (called *cells*) is a *subdivision* of $\mathcal{M}$ if it satisfies:

(a) For every cell $\sigma \in S$ the restriction $\mathcal{M}(\sigma)$ has rank r.
(b) For every one-element extension $\mathcal{M} \cup p$ of $\mathcal{M}$ and every $\sigma_1, \sigma_2 \in S$,

$$p \in conv_{\mathcal{M}\cup p}(\sigma_1) \cap conv_{\mathcal{M}\cup p}(\sigma_2) \qquad \Longrightarrow \qquad p \in conv_{\mathcal{M}\cup p}(\sigma_1 \cap \sigma_2)$$

(c) If $\sigma_1, \sigma_2 \in S$, then $\sigma_1 \cap \sigma_2$ is a common face of the two restrictions $\mathcal{M}(\sigma_1)$ and $\mathcal{M}(\sigma_2)$.

(d) If $\sigma \in S$, then each facet of $\mathcal{M}(\sigma)$ is either contained in a facet of $\mathcal{M}$ or contained in precisely two cells of S.

If all the cells are r-simplices, Definition 4.5 specializes to Definition 2.2 of an oriented matroid triangulation. Indeed, conditions (a) and (c) are then redundant and the other two are respectively our *pseudo-manifold* and *proper intersection* properties. The following results are proved in [**11**, Section 9.6] for the acyclic case, and generalize to the general case with exactly the same proofs.

- Lifting subdivisions are a particular case of a subdivision.
- If $\mathcal{M}$ is acyclic and realized by a point configuration $\mathcal{A} = \{x_1, \dots, x_n\}$ in $\mathbb{R}^{r-1}$, then a collection S of cells in $\mathcal{M}$ is a subdivision of $\mathcal{M}$ if and only if $\{conv(\sigma) : \sigma \in S\}$ is a subdivision of the polytope $P = conv(\mathcal{A})$ in the geometric sense.
- If $\mathcal{M}$ is realized by a vector configuration $\mathcal{V} = \{x_1, \dots, x_n\}$ in $\mathbb{R}^r$, then a collection S of cells in $\mathcal{M}$ is a subdivision of $\mathcal{M}$ if and only if $\{pos(\sigma) : \sigma \in S\}$ is a polyhedral fan with support $P = pos(\mathcal{V})$. I.e. a *subdivision* $\mathcal{V}$ in the sense of [**7**].

It is reasonable to think that suitable translations of the characterizations of triangulations in Theorem 2.4 yield characterizations of oriented matroid subdivisions. We will not show this. However, the following generalization of Propositions 3.5 and 4.2 in [**1**], whose proof is similar to that of our Lemma 2.8, will be of use to us.

LEMMA 4.6. *Let S be a subdivision of an oriented matroid $\mathcal{M}$. Then,*

(i) *Any interior extension $\mathcal{M} \cup p$ in general position is covered by exactly one cell of S.*

(ii) *For any two cells σ and σ' of S there is a chain of cells $\sigma = \sigma_0, \dots, \sigma_k = \sigma'$ in S such that every two consecutive cells in the chain share a facet.*

(iii) *Any interior extension of $\mathcal{M}$ is in the relative interior of a unique face of (one or several) faces of S.*

PROOF. Let r be the rank of $\mathcal{M}$. Part (b) of Definition 4.5 clearly implies that every extension $\mathcal{M} \cup p$ in general position is covered at most once in S (since the intersection of two different cells has rank strictly less than r and cannot cover p). Let $\sigma = \{a_1, \dots, a_m\} \in S$ be an arbitrary cell and let $\mathcal{M} \cup p_1$ be the lexicographic extension defined by $p_1 := [a_1^+, \dots, a_m^+]$, which is covered by σ and in general position.

We will work in the two-element extension $\mathcal{M}' = \mathcal{M} \cup \{p_1, p_2\}$ of $\mathcal{M}$, defined as the extension of $\mathcal{M} \cup p_2$ by the lexicographic expression $p_1 := [a_1^+, \dots, a_m^+]$. It is easy to check that S is also a subdivision of $\mathcal{M}'$: properties (a) and (c) of Definition 4.5 are trivial; property (b) follows from the fact that any extension of $\mathcal{M}'$ restricts to an extension of $\mathcal{M}$ and property (d) follows from the fact that p_1 and p_2 are interior and in general position and thus the proper faces of $\mathcal{M}$ and of $\mathcal{M} \cup \{p_1, p_2\}$ are the same.

We recall the following properties of $\mathcal{M}'$, used in Lemma 2.8: counting how many cells of S cover p_1 (resp. p_2) gives the same result in $\mathcal{M}'$ and in $\mathcal{M} \cup p_1$ (resp $\mathcal{M} \cup p_2$) and any vector of $\mathcal{M}'$ containing p_1 or p_2 (or both) in its support is a spanning set. As we did in the proof of Lemma 2.8, we consider the following directed graph $G_{[p_1, p_2]}$ whose nodes are some of the cells of S:

- a cell $\sigma \in S$ is a node in the graph if and only if $(\{p_1, p_2\}, \sigma)$ is a vector of $\mathcal{M}'$.

- let τ be a certain $(r-1)$-face of a simplex of S for which $(\{p_1, p_2\}, \tau)$ is a vector of $\mathcal{M}'$. In particular, τ is not in a facet of $\mathcal{M}$ and there are exactly two cells σ^+ and σ^- in S having τ as a facet. Let $C = (C^+, C^-)$ be the cocircuit of $\mathcal{M}'$ vanishing on τ, and assume without loss of generality that $p_1 \cup (\sigma^+ \setminus \tau) \subseteq C^+$ and $p_2 \cup (\sigma^- \setminus \tau) \subseteq C^-$. Then, introduce a directed edge going from σ^+ to σ^-.

We claim that the connected components of the graph $G_{[p_1,p_2]}$ obtained in this way are either isolated points, or linear paths coherently oriented, or oriented cycles. We also claim that the isolated points correspond to cells containing both p_1 and p_2 in the convex hull and that the starting and end points of the linear paths correspond, respectively, to cells of S having p_1 and p_2 (but not both) in the convex hull. These claims imply that p_2 is covered by the same number (equal to 1) of cells as $\mathcal{M} \cup p_2$. This implies part (i). Part (ii) follows by taking p_1 and p_2 to be lexicographic extensions in the relative interior of any two specific cells of S. The existence in part (iii) follows by perturbing p into general and interior position, for example via a positive lexicographic perturbation. The uniqueness holds by property (b) in Definition 4.5.

The claims follow from the following facts:

(1) if a cell σ has $p_1 \in conv_{\mathcal{M}'}(\sigma)$ and $p_2 \notin conv_{\mathcal{M}'}(\sigma)$, then it is a node of the graph and there is a unique edge incident to it, which is out-going.

(2) if a cell σ has $p_1 \notin conv_{\mathcal{M}'}(\sigma)$ and $p_2 \in conv_{\mathcal{M}'}(\sigma)$, then it is a node of the graph and there is a unique edge incident to it, which is in-going.

(3) any other cell σ which is a node in the graph $G_{[p_1,p_2]}$ is either an isolated node or has two edges incident to it, one in-going and one out-going.

In Lemma 2.8 we proved these facts using realizability of corank 2 oriented matroids. Here we need to use a different proof. Observe that since the two extensions are in general position, they being in the convex hull of a cell is the same as being in the relative interior.

Let us prove (1). Thus, we assume p_1 to be in the relative interior of σ and p_2 not to be. We can assume σ not to be totally cyclic: otherwise it would be the only cell of S. If σ is not totally cyclic, then the oriented matroid $\widehat{\mathcal{M}_0} := \mathcal{M}'(\sigma \cup \{p_1\})$ is a non-cyclic lift of $\mathcal{M}_0 := \mathcal{M}'(\sigma \cup \{p_1\})/p_1$. Let S_0 be the lifting subdivision of $\mathcal{M}_0$ induced. By inductive hypothesis, the interior extension in general position $\mathcal{M}_0 \cup p_2 := \mathcal{M}'(\sigma \cup \{p_1, p_2\})/p_1$ has the extension element p_2 contained in precisely one cell of S_0, which means that there is exactly one facet τ of σ such that $(\tau, \{p_2\})$ is a vector of $\mathcal{M}_0 \cup p_2$. The fact that p_1 is in the relative interior of σ and p_2 is outside its convex hull implies that this vector extends to the vector $(\tau, \{p_1, p_2\})$, and that the corresponding edge of $G_{[p_1,p_2]}$ is the unique edge incident to the node of σ and is oriented as desired. The proof of (2) is completely analogue.

We finally deal with (3). We assume σ to be a cell of S which gives a node in the graph with at least one edge, but not in the conditions of (1) or (2). Thus, $(\sigma, \{p_1, p_2\})$ is a vector. Also, there is a facet τ of σ such that $(\tau, \{p_1, p_2\})$ is a vector. The latter implies that one of p_1 and p_2 is not in the convex hull of σ. In order not to be in cases (1) or (2), the other one must also not be in the convex hull of σ.

As before, consider the oriented matroid $\widehat{\mathcal{M}_0} := \mathcal{M}'(\sigma \cup p_1)$. Observe that p_1 cannot be in a positive circuit of $\widehat{\mathcal{M}_0}$. Otherwise, elimination of p_1 in this circuit and the vector $(\sigma, \{p_1, p_2\})$ would imply that p_2 is in the convex hull of σ, in $\mathcal{M} \cup \{p_1, p_2\}$. Hence, $\widehat{\mathcal{M}_0}$ is a non-cyclic lift of $\mathcal{M}_0 := \mathcal{M}'(\sigma \cup \{p_1\})/p_1$.

Thus, let S_0 be the lifting triangulation of $\mathcal{M}_0$ associated to the lift $\widehat{\mathcal{M}_0}$. The inductive argument shows that the extension extension $\mathcal{M}_0 \cup p_2 = \mathcal{M}'(\sigma \cup \{p_1, p_2\})/p_1$ of $\mathcal{M}_0$ is in a unique cell of the lifting subdivision, i.e. there is a unique facet τ_1 of σ such that the cocircuit $C_1 = (C_1^+, C_1^-)$ vanishing on τ_1 has $p_1 \in C_1^+$, $p_2 \in C_1^-$ and $\sigma \cap C_1^- = \emptyset$. With the same arguments applied to p_2, we obtain a unique facet τ_2 such that the cocircuit $C_2 = (C_2^+, C_2^-)$ vanishing on it has $p_1 \in C_2^+$, $p_2 \in C_2^-$ and $\sigma \cap C_2^+ = \emptyset$. These two facets cannot coincide, (otherwise they would span σ) and thus provide the unique two edges incident to the vertex of $G_{[p_1,p_2]}$ corresponding to σ. The edge corresponding to τ_1 is out-going and the one corresponding to τ_2 is in-going. □

COROLLARY 4.7.
(i) *Let S be a subdivision of an oriented matroid $\mathcal{M}$ and let $\mathcal{M} \cup p$ be an interior extension of $\mathcal{M}$. Then, there is at least one cell $\sigma \in S$ with $p \in conv_{\mathcal{M} \cup p}(\sigma)$.*
(ii) *Let S and S' be two subdivisions of an oriented matroid $\mathcal{M}$. If one is contained in the other, then they coincide.*

PROOF. Part (i) is straightforward from parts (i) and (iii) of Lemma 4.6.

For (ii), suppose that $S' \subseteq S$ and that there is a cell $\sigma \in S \setminus S'$. We consider any extension $\mathcal{M} \cup p$ of $\mathcal{M}$ in general position and in the convex hull of σ (such as a positive lexicographic extension by the elements in σ). By Lemma 4.6 there is a cell $\sigma' \in S'$ with p in the convex hull of σ', but this is a contradiction with the same lemma, since both σ and σ' are cells in S. □

The question of how to recover the lift/extension associated to a lifting subdivision of an oriented matroid is answered in the following lemma. The answer is partial, because different lifts can produce the same subdivision.

LEMMA 4.8. *Let S be a lifting subdivision of an oriented matroid $\mathcal{M}$, defined by the acyclic lift $\widehat{\mathcal{M}}$. Let $\mathcal{M}^* \cup \overline{p}$ be the extension of $\mathcal{M}^*$ obtained by reorientation of p in the totally cyclic extension $\mathcal{M}^* \cup p$ dual to $\widehat{\mathcal{M}}$. Then, the following properties hold for every cocircuit $C = (C^+, C^-)$ of $\mathcal{M}^*$:*

- *If some cell of S contains the support $\underline{C} = C^+ \cup C^-$ of C then the cocircuit signature of the extension $\mathcal{M}^* \cup \overline{p}$ at the cocircuit C of $\mathcal{M}^*$ is $C(\overline{p}) = 0$.*
- *Otherwise, if some cell of S contains C^+ (resp. C^-), then $C(\overline{p}) = -1$ (resp. $C(\overline{p}) = +1$).*

PROOF. Observe that for any circuit $C = (C^+, C^-)$ of $\mathcal{M}$ exactly one of $(C^+ \cup \{\widehat{p}\}, C^-)$ $(C^+, C^- \cup \{\widehat{p}\})$ and (C^+, C^-) is a circuit of the lift $\widehat{\mathcal{M}}$, where $\widehat{p}$ denotes the extra element in the lift. If a cell of S contains $\underline{C}$, then $\underline{C}$ is in a facet of $\widehat{\mathcal{M}}$ not containing $\widehat{p}$. This implies that $rank(\underline{C} \cup \{\overline{p}\}) = rank(\underline{C}) + 1$ and the only possible circuit of the three above is (C^+, C^-) itself. Thus, $C = (C^+, C^-)$ is a cocircuit of $\mathcal{M}^* \cup \overline{p}$; that is, $C(\overline{p}) = 0$. If no cell contains $C^+ \cup C^-$, assume that a cell of S contains C^+ but not C^- (the other case is analogue). Then there is a positive cocircuit D of $\widehat{\mathcal{M}}$ which has empty intersection with C^+ but not empty with C^-. Orthogonality of circuits and cocircuits implies that the only possible extension of the circuit C to $\widehat{\mathcal{M}}$ is $(C^+ \cup \widehat{p}, C^-)$, which is as required in order that the cocircuit signature of $\mathcal{M}^* \cup \overline{p}$ at C be $C(\overline{p}) = -1$. □

4.2. Lawrence polytopes only have lifting subdivisions

We recall that an oriented matroid $\mathcal{M}$ is called *polytopal* (or a *matroid polytope*) if every one-element subset is a face. In particular, $\mathcal{M}$ has to be acyclic. Every oriented matroid $\mathcal{M}$ of rank r on n elements has an associated oriented matroid $\Lambda(\mathcal{M})$ of rank $n+r$ on $2n$ elements which is (essentially) polytopal and has the property that the whole oriented matroid structure of $\mathcal{M}$ is contained in the Las Vergnas face lattice of $\Lambda(\mathcal{M})$, i.e. in the structure of the boundary of $\Lambda(\mathcal{M})$. This construction was invented by Jim Lawrence (unpublished). References for the construction are [**5**], [**10**], [**11**, Section 9.3], [**38**, Chapter 7] and [**40**, p. 180]; the latter two deal only with the realizable case.

In this section we will see that every subdivision of the *Lawrence polytope* $\Lambda(\mathcal{M})$ is a lifting subdivision and that, in fact, there is a 1-to-1 correspondence between the subdivisions of $\Lambda(\mathcal{M})$ and the extensions (interior or not) of the dual oriented matroid $\mathcal{M}^*$ of $\mathcal{M}$. Under this correspondence flips correspond exactly to mutations. The correspondence will have the interesting consequence of relating the *extension space conjecture* to a conjecture regarding subdivisions of polytopes. This will be shown in Section 4.3.

Let us fix some notation. Let $\mathcal{M}$ be an oriented matroid of rank r on a set E of n elements and let $\mathcal{M}^*$ be its dual. We construct an oriented matroid $\Lambda(\mathcal{M})^*$ on the set $E\times\{1,-1\}$, which can be geometrically interpreted (e.g., in a realized setting) as the union of $\mathcal{M}^*$ and its image by the central inversion. We identify E with $E\times\{1\}$ and will write $\overline{A}$ to denote $A\times\{-1\}$, for every subset A of E. Then, $\Lambda(\mathcal{M})^*$ is the extension of $\mathcal{M}$ by n elements $\overline{e}\in\overline{E}$ antiparallel to the corresponding $e\in E$. In other words, $\Lambda(\mathcal{M})^*$ is characterized by:

- $\Lambda(\mathcal{M})^*$ is an oriented matroid of rank $n-r$ on $E\cup\overline{E}$ whose restriction to E is $\mathcal{M}^*$.

- For any $e\in E$, the element $\overline{e}\in\overline{E}$ is a loop in $\Lambda(\mathcal{M})^*$ if and only if $e\in E$ is a loop and, if it is not a loop, then $(\{e,\overline{e}\},\emptyset)$ is a positive circuit of $\Lambda(\mathcal{M})^*$.

Let $\Lambda(\mathcal{M})$ be the dual of $\Lambda(\mathcal{M})^*$. From the construction it follows that $(\{e,\overline{e}\},\emptyset)$ is a covector of $\Lambda(\mathcal{M})$ for every $e\in E$ and, in particular, that $\{e,\overline{e}\}$ is a face (and the complement of a face as well). If $\{e,\overline{e}\}$ has rank 2, then both e and $\overline{e}$ are vertices. If the rank is 1, then e and $\overline{e}$ are parallel elements and they form a "double" vertex of $\Lambda(\mathcal{M})$. This happens if and only if e is a loop in $\mathcal{M}$.

Therefore, $\Lambda(\mathcal{M})$ is polytopal if $\mathcal{M}$ is loop-less, and "almost" polytopal otherwise. If $\mathcal{M}$ is realized by the column vectors of a matrix M, then $\Lambda(\mathcal{M})$ is realized by the column vectors of

$$\Lambda(M) := \begin{pmatrix} M & \mathbf{0} \\ I & I \end{pmatrix},$$

where I and $\mathbf{0}$ are the identity and zero matrices of the appropriate sizes ([**5**], [**38**, Chapter 7]). A different description of $\Lambda(\mathcal{M})$ in the realizable case is by a sequence of *Lawrence extensions* (see [**40**, Theorem and Definition 6.26]).

Definition 4.9. In the above conditions we say that $\Lambda(\mathcal{M})$ is the *Lawrence matroid polytope* (or Lawrence polytope, for short) associated to $\mathcal{M}$.

A basic property relating $\mathcal{M}^*$ and $\Lambda(\mathcal{M})^*$ is that there is a canonical bijection between their lattices of covectors (in a pseudo-sphere arrangement representation, $\Lambda(\mathcal{M})^*$ is obtained from $\mathcal{M}^*$ by considering each hypersphere twice, one with each orientation). This implies a canonical bijection between their extensions and that every extension of $\mathcal{M}^*$ is interior when regarded in $\Lambda(\mathcal{M})^*$ since $\Lambda(\mathcal{M})^*$ is totally cyclic. Hence, lifting subdivisions of $\Lambda(\mathcal{M})$ are defined by extensions of $\mathcal{M}^*$, interior or not.

Let us now characterize the circuits, cocircuits and bases of a Lawrence polytope. We introduce the following notation.

DEFINITION 4.10. Let B be a subset of E, and $A \subseteq B$. We denote by

$$_AB := (B \setminus A) \cup \overline{A},$$

and call it the *reorientation of B at A.*

Let $C = (C^+, C^-)$ be a signed subset of E and let $A \subseteq C^+ \cup C^-$. We denote by

$$_AC = (\ (C^+ \setminus A) \cup \overline{(C^- \cap A)}\quad ,\quad (C^- \setminus A) \cup \overline{(C^+ \cap A)}\ ),$$

and call it the *reorientation of C at A.*

LEMMA 4.11. *Let $\Lambda(\mathcal{M})$ be the Lawrence polytope associated with an oriented matroid $\mathcal{M}$. Let $\mathcal{C}i(\mathcal{M})$, $\mathcal{C}o(\mathcal{M})$ and $\mathcal{B}(\mathcal{M})$ denote respectively the sets of circuits, cocircuits and bases of $\mathcal{M}$. Then,*

(i) *The set of circuits of $\Lambda(\mathcal{M})$ is*

$$\mathcal{C}i_{\Lambda(\mathcal{M})} := \{(C^+ \cup \overline{C^-}, C^- \cup \overline{C^+}) : (C^+, C^-) \in \mathcal{C}i(\mathcal{M})\}.$$

(ii) *The set of cocircuits of $\Lambda(\mathcal{M})$ is*

$$\mathcal{C}o_{\Lambda(\mathcal{M})} := \{_AC : C \in \mathcal{C}o(\mathcal{M}), A \subseteq C^+ \cup C^-\} \quad \cup$$

$$\{\ (\{e, \overline{e}\}, \emptyset)\ ,\ (\emptyset, \{e, \overline{e}\}) : e \in E \text{ is not a coloop of } \mathcal{M}\}.$$

(iii) *The set of bases of $\Lambda(\mathcal{M})$ is*

$$\mathcal{B}_{\Lambda(\mathcal{M})} := \{_A(E \setminus B) \cup B \cup \overline{B} : B \in \mathcal{B}(\mathcal{M}), A \subseteq E \setminus B\}.$$

PROOF. The proof is easy via the duals $\mathcal{M}^*$ and $\Lambda(\mathcal{M})^*$ of $\mathcal{M}$ and $\Lambda(\mathcal{M})$. Parts (i) and (ii) appear in Lemma 9.3.1 and Proposition 9.3.3 of [**11**]. □

A first interesting consequence of this lemma is the fact (known in the realized case, see [**4**, pp. 310–311]) that all the triangulations of a Lawrence polytope have the same number of simplices:

PROPOSITION 4.12. *Let T be a triangulation of a Lawrence polytope $\Lambda(\mathcal{M})$. Then, for every basis B of $\mathcal{M}$ there is a unique subset $A \subseteq (E \setminus B)$ such that $_A(E \setminus B) \cup B \cup \overline{B} \in T$. In particular, all the triangulations of $\Lambda(\mathcal{M})$ have the same number of simplices, equal to the number of bases of $\mathcal{M}$.*

PROOF. Let B be a basis of $\mathcal{M}$. Then, $E \setminus B$ is a basis of $\mathcal{M}^*$ and the collection of reorientations $_A(E \setminus B)$ of $E \setminus B$ is a triangulation of $\Lambda(\mathcal{M})^*$. From part (c) of Theorem 3.8 we conclude that T has exactly one simplex which is the complement of a reorientation of $E \setminus B$. □

Now we prove that Lawrence polytopes only have lifting subdivisions. The following statement actually gives much more information.

LEMMA 4.13. *Let $\Lambda(\mathcal{M})$ be the Lawrence polytope associated to an oriented matroid $\mathcal{M}$. Let S be a subdivision of $\Lambda(\mathcal{M})$. Then:*

(i) *The support of every circuit C of $\Lambda(\mathcal{M})$ is a face of $\Lambda(\mathcal{M})$.*

(ii) *Let F be a face of $\Lambda(\mathcal{M})$ which is the support of a circuit C. Let k be the rank of F. Consider the "restriction" of S to F defined as:*

$$S_F := \{\sigma \cap F : \sigma \in S,\ rank(\sigma \cap F) = k\}.$$

Then, S_F is either the trivial subdivision $\{F\}$ of the restricted oriented matroid $\Lambda(\mathcal{M})(F)$ or one of the triangulations T_C^+ or T_C^- of a circuit introduced in Proposition 2.13.

(iii) *The cocircuit signature of $\Lambda(\mathcal{M})^*$ defined by $C(p) = 0$ if $S_{\underline{C}} = \{\underline{C}\}$ and $C(p) = +1$ (resp. $C(p) = -1$) if $S_{\underline{C}} = T_C^+$ (resp. $S_{\underline{C}} = T_C^-$) is the cocircuit signature of an extension $\Lambda(\mathcal{M})^* \cup p$ of $\Lambda(\mathcal{M})^*$.*

(iv) *S is the lifting subdivision corresponding to the acyclic lift $\widehat{\mathcal{M}}$ which is dual to the reorientation at p of the extension defined in (iii).*

PROOF. (i) By part (i) of Lemma 4.11, every circuit of $\Lambda(\mathcal{M})$ is of the form $(C^+ \cup \overline{C^-}, C^- \cup \overline{C^+})$, where (C^+, C^-) is a circuit of $\mathcal{M}$. In the other hand, by part (ii) of the same lemma, every set of the form $A \cup \overline{A}$ is a face of $\Lambda(\mathcal{M})$.

(ii) Let $C = (C^+ \cup \overline{C^-}, C^- \cup \overline{C^+})$ be the circuit whose support equals F. Let $a_1 \in C^+ \cup \overline{C^-}$, $a_2 \in C^- \cup \overline{C^+}$. We first prove that at most one of F, $F \setminus \{a_1\}$ or $F \setminus \{a_2\}$ lies in S_F. Let $F \setminus \{a_1, a_2\} = \{b_1, \ldots, b_{k-1}\}$. Consider the lexicographic extension $\Lambda(\mathcal{M}) \cup p_F$ of $\Lambda(M)$ defined by the expression $p_F := [b_1^+, \ldots, b_{k-1}^+, a_2^+]$. It lies both in the relative interiors of $F \setminus \{a_1\}$ and $F \setminus \{a_2\}$ (the first thing is trivial, the second follows from the fact that a_1 and a_2 lie in opposite parts of the circuit). Then, condition (b) in Definition 4.5 applied to S implies that only one of $F \setminus \{a_1\}$ or $F \setminus \{a_2\}$ can be the intersection with F of a cell of S. If one of them is, then condition (c) in the same definition implies that F cannot be contained in a cell of S.

Thus, for any pair of elements $a_1 \in C^+ \cup \overline{C^-}$ and $a_2 \in C^- \cup \overline{C^+}$ at most one of F, $F \setminus \{a_1\}$ or $F \setminus \{a_2\}$ is in S_F. Since any spanning subset of F is either F or of the form $F \setminus \{a\}$ (F is the support of a circuit), S_F is contained in one of the three subdivisions $\{F\}$, T_C^+ or T_C^- of F. If $S_F \subseteq \{F\}$ then clearly $\{F\} = S_F$. Otherwise, suppose without loss of generality that S_F is contained in T_C^-. We have to prove that then $S_F = T_C^-$. If this is not the case, suppose that $F \setminus \{a_2\} \in T_C^-$ is one of the missing simplices. This is impossible, because then the lexicographic extension p_F defined above would not lie in the convex hull of any cell of S_F and, thus, would not lie in the convex hull of any cell of S, which contradicts part (i) of Corollary 4.7.

(iii) To prove that the cocircuit signature defines an extension it suffices to show that it defines an extension on every rank 2 contraction of $\Lambda(\mathcal{M})^*$, by Lemma 1.3.

If e is a non-loop of such a contraction, then $\overline{e}$ is also a non-loop, and vice versa. Thus, we can assume the contraction to be $\Lambda(\mathcal{M})^*/(A \cup \overline{A})$. The dual of the contraction is the restriction of $\Lambda(\mathcal{M})$ to a corank 2 face $(E \setminus A) \cup (\overline{E} \setminus \overline{A})$ of $\Lambda(\mathcal{M})$.

Observe that the cocircuit signature restricted to $\Lambda(\mathcal{M})^*/(A \cup \overline{A})$ can be obtained from the subdivision S restricted to $\Lambda(\mathcal{M}) \setminus (A \cup \overline{A})$ in the same way as we obtained the cocircuit signature for p from S. Since a corank 2 oriented matroid is always realizable and every subdivision of an acyclic realized oriented matroid of corank 2 is regular (in particular, lifting) by [**24**, Theorem 4] (see also Proposition 5.8 in [**14**]), we conclude that the restriction of the cocircuit signature to every rank 2 contraction is the cocircuit signature of a lifting triangulation. Thus, the cocircuit signature for p in $\Lambda(\mathcal{M})^*$ defines an extension $\Lambda(\mathcal{M})^* \cup p$.

(iv) Let S' be the subdivision of $\Lambda(\mathcal{M})$ defined by that lift. We have to prove that $S = S'$; by part (ii) of Corollary 4.7 it is enough to prove that any cell of S is a cell of S' as well.

Let σ be a cell of S and let σ^c denote its complement in the set of elements of $\Lambda(\mathcal{M})$. We check the following two properties, the first of which is trivial:

(a) If a cocircuit C of $\Lambda(\mathcal{M})^*$ has support contained in σ, then $C(p) = 0$.

(b) For any element $a \in \sigma^c$, there is a cocircuit $C_a = (C_a^+, C_a^-)$ of $\Lambda(\mathcal{M})^*$ with $C_a(p) = +1$, $C_a^+ \cap \sigma^c = \{a\}$ and $C_a^- \cap \sigma^c = \emptyset$. Indeed, since σ is spanning in $\Lambda(\mathcal{M})$, there is a circuit C of $\Lambda(\mathcal{M})$ with support containing a and contained in $\sigma \cup \{a\}$, which we assume to be positive at a. Let $\underline{C}$ be the support of C. Since $\sigma \cap \underline{C} = \underline{C} \setminus \{a\}$, part (ii) of the Lemma implies that T_C^+ is a subcomplex of S, that is, $C(p) = +1$.

Statement (a) implies that p is in the flat spanned by σ^c; that is, that $\sigma^c \cup \{p\}$ contains the support of a circuit. The only way in which this circuit can be orthogonal to the cocircuits in (b) is that it equals $(\{p\}, \sigma^c)$ (or its opposite). Hence, $(\emptyset, \sigma^c \cup \{\widehat{p}\})$ is a cocircuit of the lift $\widehat{\Lambda(\mathcal{M})}$ defining S' and σ is a cell of S'. □

THEOREM 4.14. *Let $\Lambda(\mathcal{M})$ be the Lawrence polytope associated to an oriented matroid $\mathcal{M}$. Then:*

(i) *There is a natural bijection between the extensions of $\mathcal{M}^*$ and the subdivisions of $\Lambda(\mathcal{M})$.*
(ii) *Under this bijection, two triangulations differ by a geometric bistellar flip if and only if the corresponding extensions differ by a mutation.*

PROOF. (i) Definition 4.1 provides a natural map from the collection of interior extensions of $\Lambda(\mathcal{M})^*$ to the subdivisions of $\Lambda(\mathcal{M})$. Since $\Lambda(\mathcal{M})^*$ is totally cyclic, all its extensions are interior and by construction of $\Lambda(\mathcal{M})^*$ they are in bijection with the extensions (interior or not) of $\mathcal{M}^*$. Thus, we have a natural map from the extensions of $\mathcal{M}^*$ to the subdivisions of $\Lambda(\mathcal{M})$. Since all the subdivisions of $\Lambda(\mathcal{M})$ are lifting subdivisions, the map is surjective. The fact that the complete cocircuit signature of an extension can be recovered from the corresponding lifting subdivision (parts (iii) and (iv) of the previous lemma), implies that the map is injective.

(ii) If two extensions of $\mathcal{M}^*$ in general position differ by a mutation, Theorem 3.14 implies that the associated triangulations of $\Lambda(\mathcal{M})$ differ by a flip (since they cannot be equal).

Reciprocally, suppose that T_1 and T_2 are two triangulations of $\Lambda(\mathcal{M})$ differing by a flip. Let C be the circuit of $\Lambda(\mathcal{M})$ in which the flip is supported, which is a cocircuit of $\Lambda(\mathcal{M})^*$. Let $\Lambda(\mathcal{M})^* \cup p_1$ and $\Lambda(\mathcal{M})^* \cup p_2$ be the extensions of $\Lambda(\mathcal{M})^*$ in general position corresponding to the triangulations T_1 and T_2. We have that $Ext_{p_1} - Ext_{p_2}$ is a sum of cocircuit vectors supported on C, by Definition 3.11 and Corollary 3.3. We want to derive from this that the two extensions differ by a mutation, that is, that the cocircuit signatures of p_1 and p_2 differ only on the cocircuit C (and its opposite).

Let C' be a cocircuit of $\Lambda(\mathcal{M})^*$ not equal to C or its opposite. In particular, the hyperplanes H_C and $H_{C'}$ in which C and C' respectively vanish do not coincide. Let us prove that $C'(p_1) = C'(p_2)$. For this let σ be any full rank simplex with a facet contained in $H_{C'}$. Since all the reorientations of σ form a triangulation, there is at least one such reorientation having p_1 in its convex hull. Let this reorientation be $\tau \cup \{a\}$, with $\tau \subseteq H_{C'}$ and hence $C'(a) = C'(p_1) \in \{+1, -1\}$

If also $p_2 \in conv_{\Lambda(\mathcal{M})^* \cup p_2}(\tau \cup a)$ we conclude that $C'(p_1) = C'(p_2)$, as we wished. If not, then $\tau \cup \{a\}$ appears in the difference $Ext_{p_1} - Ext_{p_2}$ and so it has a facet in the hyperplane H_C in which C vanishes. In particular, $a \in H_C$ and there is an element $b \in \tau$ with $\rho := \tau \setminus \{b\} \in H_C \cap H_{C'}$.

Now, T_1 and T_2 differ by a flip on the circuit (of $\Lambda(\mathcal{M})$) C, which has b and $\bar{b}$ on opposite sides. Since the complement of $\rho \cup \{a, b\}$ is in T_1 and b is the only element of $\rho \cup \{a, b\}$ in the support of C, the complement of $\rho \cup \{a, \bar{b}\}$ is in T_2 by Proposition 3.13. I.e., $\rho \cup \{a, \bar{b}\}$ appears in Ext_{p_2}. But $\rho \cup \{\bar{b}\} \subseteq H_{C'}$ again implies that $C'(p_2) = C'(a) = C'(p_1)$. □

4.3. The extension space conjecture and the Baues problem

Theorem 4.14 relates two important questions in geometric combinatorics: the conjecture that all the extensions of a realizable oriented matroid are connected by mutations and the question of whether all the triangulations of a point configuration are connected by flips. More precisely:

COROLLARY 4.15. (i) *Let $\mathcal{M}$ be an oriented matroid. Then, the collection of all the extensions of $\mathcal{M}$ in general position is connected by mutations if and only if the collection of all the triangulations of the Lawrence polytope $\Lambda(\mathcal{M}^*)$ is connected by flips.*

(ii) *There exist (non-realizable) Lawrence polytopes with triangulations which do not admit any flip. In particular, whose set of triangulations is not connected by flips.*

PROOF. Part (i) is trivial from Theorem 4.14. Part (ii) follows from the fact that there are (uniform) oriented matroids with extensions in general position which admit no mutation. The first such example was obtained by Richter-Gebert (see Theorem 2.3 of [**33**]), with rank 4 and 19 elements. Recently, J. Bokowski and H. Rohlfs [**12**] have found a smaller example with the same rank and 17 elements. This produces a Lawrence polytope with dimension 20 and 34 vertices. □

Even more, it is obvious from our results that the poset of extensions of $\mathcal{M}^*$ ordered by weak maps (or equivalently, by "perturbation") is isomorphic to the poset of subdivisions of $\Lambda(\mathcal{M})$ ordered by refinement. In the realizable case, these two posets appear as particular cases of the so-called "generalized Baues problem" (see [**32**] for a recent survey on the topic).

Given two polytopes $P \subseteq \mathbb{R}^p$ and $Q \subseteq \mathbb{R}^q$ and a projection map $\pi : \mathbb{R}^p \to \mathbb{R}^q$ with $\pi(P) = Q$, Billera et al. have defined the concept of a *subdivision of Q induced by π from P* and asked whether the order poset of these induced subdivisions ordered by refinement has always the homotopy type of a sphere of dimension $dim(P) - dim(Q) - 1$ [**8**]. The poset of π-induced subdivision is called the *Baues poset* of the projection and the question on its homotopy type is referred to as the *generalized Baues problem* since a special case where P is a hypercube and Q has dimension one appeared in a conjecture of Baues, solved in [**8**].

The generalized Baues problem has been given a negative answer by Rambau and Ziegler (see [**31**]) with a counterexample in which $\dim(P) = 5$, $\dim(Q) = 2$

and P has 10 vertices. The counterexample is minimal in dimension of both P and Q since the Baues poset is known to have the homotopy type of a sphere if $\dim(Q) = 1$ [**8**] or $\dim(P) - \dim(Q) \leq 2$ [**31**].

Specially interesting are the following two cases of the generalized Baues problem:

(i) If P is a simplex, then the π-induced subdivisions of Q are all the polytopal subdivisions of Q which use (perhaps not all of) the image points by π of the vertices of P. The Baues problem asks whether the order poset of all proper subdivisions of the point configuration $\pi(vert(P))$ has the homotopy type of a sphere. This is known to be true if he point configuration has dimension at most 2 [**16**] or if it has at most 4 points more than its dimension [**3**]. Very recently, the author has found a triangulation without geometric bistellar flips of a point configuration in dimension 6 with 324 points [**36**]. The cases of dimension between 3 and 5, and the case of general position (meaning uniform oriented matroid) remain open.

(ii) If P is a hypercube then Q is a zonotope, that is, the Minkowski sum of several segments (see [**40**, Section 7.3]). In this case, the π-induced subdivisions of Q coincide with the so-called *zonotopal subdivisions* (see [**11**, page 60]) or *zonotopal tilings* (see [**40**, Section 7.5]) of the zonotope Q. The Bohne-Dress Theorem on zonotopes (see [**11**, Theorem 2.2.13] or [**40**, Theorem 7.32]) implies that the zonotopal subdivisions of Q are in poset isomorphism with the lifts of the associated oriented matroid, i.e. with extensions of the dual, with refinement of subdivisions corresponding to perturbation of extensions. That is, the zonotopal case of the Generalized Baues problem is equivalent to the *extension space conjecture* stating that the extension space of a realizable oriented matroid $\mathcal{M}$ of rank r has the homotopy type of a $(r-1)$-sphere. The cases of rank at most 3 or corank at most 2 are answered positively in [**39**].

Theorem 4.14 implies the equivalence of (i) and (ii) in the following statement. The equivalence of (i) and (iii) is the afore-mentioned Bohne-Dress Theorem on zonotopes. Let us mention that [**21**] contains a direct, geometric proof of the equivalence between (ii) and (iii).

COROLLARY 4.16. *Let $\mathcal{M}$ be an oriented matroid of rank r with n elements. Then the following three posets are isomorphic:*

(i) *The poset of all the extensions of $\mathcal{M}^*$, ordered by weak maps.*
(ii) *The poset of all the subdivisions of the Lawrence polytope $\Lambda(\mathcal{M})$, ordered by refinement.*
(iii) *If $\mathcal{M}$ is realized by a vector configuration V, the poset of all the zonotopal tilings of the zonotope $Z(V)$ generated by V.*

In particular, the extension space conjecture is equivalent to the following one: for any realized Lawrence polytope Λ of dimension d with n vertices, the poset of proper polytopal subdivisions of Λ has the homotopy type of a $(n-d-2)$-sphere. □

EXAMPLE 4.17. The above statement can be easily checked in the following simple example, in which all the triangulations/extensions which appear are lexicographic.

Let $\mathcal{M}$ be an oriented matroid of rank 1 and with n elements, none of them loops. Up to reorientation, $\mathcal{M}$ can be realized by the vector $v = (1, 1, \ldots, 1)$ with n entries all equal to 1. The zonotope Z associated to $\mathcal{M}$ is the Minkowski sum of n segments in the same direction. The zonotopal tilings of Z can be thought of as monotone strings of faces of an n-cube, and the poset of all them is the face lattice of an $(n-1)$-permutahedron [**40**, pages 301–304]. We recall that the permutahedron is the polytope of dimension $n-1$ in $\mathbb{R}^n$ whose vertices are the points obtained by permuting in all the possible ways the entries of the point $(1, 2, \ldots, n)$ [**40**, page 17].

The Lawrence polytope $\Lambda(\mathcal{M})$ can be realized as the oriented matroid of affine dependencies of the product of a segment and an $(n-1)$-simplex. It is well-known (cf. for example [**19**, pages 243–246]) that the poset of subdivisions of this polytope is again isomorphic to the face lattice of the permutahedron.

The dual $\mathcal{M}^*$ of $\mathcal{M}$ is the unique uniform totally cyclic oriented matroid of rank $n-1$ with n elements, which can be realized by the vectors joining the barycenter of an $(n-1)$-simplex to its vertices. All the extensions of $\mathcal{M}^*$ are realizable and each of them corresponds to a flag of proper faces of the $(n-1)$-simplex. Thus, the poset of extensions of $\mathcal{M}^*$ is anti-isomorphic to the face lattice of the barycentric subdivision of the boundary of an $(n-1)$-simplex; that is, isomorphic to the face lattice of the permutahedron.

4.4. A reoriented Lawrence construction

Here we introduce a reoriented version of the Lawrence construction. The construction is interesting because, applied to an acyclic non-polytopal oriented matroid $\mathcal{M}$, it produces a matroid polytope $\Sigma(\mathcal{M})$ with exactly the same collection of triangulations as $\mathcal{M}$. In other words, the "polytopal case" cannot be considered simpler than the "acyclic case" when dealing with triangulations (unless we are interested in a fixed rank). The same construction has been used in [**15**].

Let $\mathcal{M}$ be an oriented matroid of rank r on n elements which we identify with $E := \{1, \ldots, n\}$. Assign a positive integer k_i to each element i of n. Let k be the sum of these integers. We consider the oriented matroid $\Sigma(\mathcal{M})^*$ constructed from the dual $\mathcal{M}^*$ of $\mathcal{M}$ by substituting k_i copies of i for each element i. $\Sigma(\mathcal{M})^*$ has k elements and rank $n-r$. Thus, its dual $\Sigma(\mathcal{M})$ has rank $k+r-n$ and k elements. If $k_i = 2$ for every i, then $\Sigma(\mathcal{M})$ is a reorientation of the Lawrence polytope $\Lambda(\mathcal{M})$. It will be good for us to allow the full generality of arbitrary k_i's in connection to weighted unimodular configurations which will appear in Section 5.2.2.

The elements of $\Sigma(\mathcal{M})$ lie in n equivalence classes $\Sigma(1), \ldots, \Sigma(n)$ with $k_1, \ldots, k_n$ elements respectively, each class corresponding to an element of $\mathcal{M}$. Two elements e and f of $\Sigma(\mathcal{M})$ are *co-parallel* (meaning that $(\{e\}, \{f\})$ is a cocircuit) if and only if they lie in the same class or in classes corresponding to co-parallel elements of $\mathcal{M}$.

THEOREM 4.18. *Let $\mathcal{M}$ be an oriented matroid of rank r on n elements which we identify with $E := \{1, \ldots, n\}$. Let $k_1, \ldots, k_n$ be positive integers. The oriented matroid $\Sigma(\mathcal{M})$ of rank $k+r-n$ on k elements just defined has the following properties.*

(i) *$\Sigma(\mathcal{M})$ is acyclic if and only if $\mathcal{M}$ is acyclic. If this is the case, then an element of $\Sigma(\mathcal{M})$ is a vertex (face of rank 1) if and only if the corresponding element i of $\mathcal{M}$ was a vertex of $\mathcal{M}$, or if k_i is greater than 1. In* *particular, $\Sigma(\mathcal{M})$ is polytopal if and only if $\mathcal{M}$ is acyclic and $k_i > 1$ for every non-vertex element of $\mathcal{M}$.*

(ii) *There is a natural bijective correspondence between the triangulations of $\mathcal{M}$ and the triangulations of $\Sigma(\mathcal{M})$ which preserves the features of being lifting or lexicographic.*

Although we will neither prove nor use this, the correspondence mentioned above preserves geometric bistellar flips and extends to an isomorphism of the posets of subdivisions of $\mathcal{M}$ and $\Sigma(\mathcal{M})$.

Before going into the proof, the following simple example may help to clarify the construction. Suppose that $\mathcal{A}$ is the point configuration in the plane consisting of the vertices of a convex polygon P plus an interior point p. We are going to show the construction $\Sigma(\mathcal{A})$ applied to $\mathcal{A}$ with all the parameters k_i equal to 1 except the one of the interior point p which will be equal to 2. The resulting configuration $\Sigma(A)$ in $\mathbb{R}^3$ consists of the vertices of a bipyramid, with the equator of the bipyramid being the polygon P and in such a way that the intersection of the axis with the equatorial plane of the bipyramid coincides with the point p. The reader should try to visualize in this example the correspondence between triangulations of $\mathcal{A}$ and of $\Sigma(\mathcal{A})$ exhibited in the following proof.

PROOF. (i) $\Sigma(\mathcal{M})^*$ is totally cyclic if and only if $\mathcal{M}^*$ is totally cyclic, which proves the first part of (i). For the second part, assume that $\Sigma(\mathcal{M})^*$ and $\mathcal{M}^*$ are totally cyclic. An element of an acyclic oriented matroid is a vertex if and only if the contraction at this element is acyclic. Thus, an element of $\Sigma(\mathcal{M})$ is a vertex if and only if its deletion in $\Sigma(\mathcal{M})^*$ is totally cyclic. This happens if and only if its equivalence class has at least another element or the deletion of the corresponding element of $\mathcal{M}^*$ is totally cyclic.

(ii) It is obvious how to relate lifting and lexicographic triangulations of $\mathcal{M}$ and $\Sigma(\mathcal{M})$, since the extensions of $\mathcal{M}^*$ and $\Sigma(\mathcal{M})^*$ are "the same" and an extension is interior (resp. in general position) in $\mathcal{M}^*$ if and only if it is as well in $\Sigma(\mathcal{M})^*$. Studying what this correspondence between extensions of $\mathcal{M}^*$ and $\Sigma(\mathcal{M})^*$ looks like in terms of the extension vectors we conclude the following heuristic rule for obtaining a triangulation $\Sigma(T)$ of $\Sigma(\mathcal{M})$ from a triangulation T of $\mathcal{M}$. Let $\Sigma(i)$ denote the equivalence class in the set of elements of $\Sigma(\mathcal{M})$ of the element $i \in E$. For each subset $\sigma \in E$ define the following collection of subsets of the elements of $\Sigma(\mathcal{M})$:

$$\Sigma(\sigma) = \{S : \#(S \cap \Sigma(i)) = k_i - 1 + \#(\sigma \cap \{i\})\}.$$

In other words, the subsets S appearing in $\Sigma(\sigma)$ are those which contain the equivalence classes associated to the elements of σ and miss exactly one element from the other equivalence classes. The complements of the so-defined sets S are independent (resp. spanning) in $\Sigma(\mathcal{M})^*$ if and only if σ is independent (resp. spanning) in $\mathcal{M}^*$. Thus, if T is a triangulation of $\mathcal{M}$ the following is a collection of maximal simplices of $\Sigma(\mathcal{M})$:

$$\Sigma(T) = \cup_{\sigma \in T} \Sigma(\sigma).$$

It follows from the definition of $\Sigma(T)$ that a triangulation T of $\mathcal{M}$ is the lifting triangulation corresponding to an extension $\mathcal{M}^* \cup p$ if and only if $\Sigma(T)$ is the lifting triangulation corresponding to "the same" extension $\Sigma(\mathcal{M})^* \cup p$ of $\Sigma(\mathcal{M})^*$. This proves that the correspondence $T \to \Sigma(T)$ restricts to a bijection between lifting (resp. lexicographic) triangulations of $\mathcal{M}$ and $\Sigma(\mathcal{M})$.

We will show that the correspondence is bijective on the set of all triangulations of $\mathcal{M}$ and $\Sigma(\mathcal{M})$ using the characterization of triangulations which appears in Theorem 3.8. If e and f are elements of $\Sigma(\mathcal{M})^*$ in the equivalence class of a non-loop element of $\mathcal{M}$, then the signed set $(\{f\}, \{e\})$ is a circuit of $\Sigma(\mathcal{M})^*$. A collection of maximal simplices of $\Sigma(\mathcal{M})$ satisfies the circuit equations of Theorem 3.8 for all the circuits of this type if and only if it is a union of collections of simplices of the form $\Sigma(\sigma)$ for different maximal simplices σ of $\mathcal{M}$.

The rest of the circuits of $\Sigma(\mathcal{M})^*$ are obtained from circuits of $\mathcal{M}^*$ by choosing a representative in $\Sigma(\mathcal{M})^*$ for each element of $\mathcal{M}$. Thus, a collection T of simplices of $\mathcal{M}$ satisfies the circuit equations of Theorem 3.8 if and only $\Sigma(T)$ satisfies the circuit equations as well.

Finally, it is obvious that the triangulations of $\Sigma(\mathcal{M})^*$ are all obtained from the triangulations of $\mathcal{M}^*$ by choosing a representative of each equivalence class of elements. This implies that a collection $\Sigma(T)$ of simplices of $\Sigma(\mathcal{M})$ obtained as the union of the $\Sigma(\sigma)$ corresponding to a collection T of simplices of $\mathcal{M}$ satisfies the duality equations $\langle v_{\Sigma(T)}, v\rangle = 1$ for all the incidence vectors v of triangulations of $\Sigma(\mathcal{M})^*$ if and only if T itself satisfies the equations for the incidence vectors of triangulations of $\mathcal{M}^*$. □

REMARK 4.19. ($\Lambda(\mathcal{M})$ and $\Sigma(\mathcal{M})$ for graphic oriented matroids).
An oriented matroid $\mathcal{M}$ is called *graphic* (see [**27**]) if it is isomorphic to the cycle oriented matroid of a directed graph G. In other words, if the elements of $\mathcal{M}$ correspond to the edges of G and the signed circuits of $\mathcal{M}$ correspond to the cycles of G, with the signing given by the orientation of the edges (see [**11**, Section 1.1]).

Graphic oriented matroids form a very restricted class, strictly contained in the so-called *binary* or *regular* oriented matroids; for example, uniform oriented matroids are graphic only if they have rank or corank at most one. For this reason, directed graphs are not normally considered a good model for oriented matroids. (The situation is quite different in matroid theory, in which many results and constructions can be interpreted in terms of graphs; see, for example, [**27**]). However, the Lawrence construction, both in its original and reoriented versions, has the following very simple interpretation for graphic oriented matroids:

- Let $\mathcal{M}$ be a graphic oriented matroid corresponding to the graph G. Then, the oriented matroid $\Lambda(\mathcal{M})$ is the graphic oriented matroid of the graph $\Lambda(G)$ obtained from G by subdividing each edge into two parts and giving opposite directions to the two parts.
- Let $\mathcal{M}$ be as above, let $1, \dots, n$ denote the edges of the graph (i.e., the elements of the oriented matroid) and let $k_1, \dots, k_n$ be positive integers. Then the oriented matroid $\Sigma(\mathcal{M})$ corresponding to this choice of k_i's is the graphic oriented matroid of the graph $\Sigma(G)$ obtained from G by subdividing each edge i into k_i new edges and giving to all of them the same direction as the old edge had.

Observe that the constructions above are consistent with the fact that $\Lambda(\mathcal{M})$ is acyclic and invariant under reorientation of $\mathcal{M}$, while $\Sigma(\mathcal{M})$ is not invariant under reorientation and is acyclic only if $\mathcal{M}$ is acyclic. The graph $\Sigma(G)$ is uniquely defined by G (and the parameters k_i), but in the graph $\Lambda(G)$ we have the choice of which of the two parts of each edge of G gets each of the two directions. However, since every cycle of $\Lambda(G)$ will contain either both or none of the two sub-edges, the graphic oriented matroid obtained is the same independently of this choice. We can choose all orientations of the sub-edges to go from the old vertex of the sub-edge to the new one. With this choice the graph $\Lambda(G)$ is bipartite.

PROPOSITION 4.20. *Let $\mathcal{M}$ be an oriented matroid of rank r on n elements. Then, the following properties are equivalent:*

(i) *$\mathcal{M}$ is graphic.*
(ii) *$\Lambda(\mathcal{M})$ is a full-rank restriction of the oriented matroid of affine dependences of the product of two simplices of dimensions $n-2$ and $r-1$.*

PROOF. We recall the following elementary facts from matroid theory: the graphic oriented matroid corresponding to a connected graph $G=(V,E)$ has $|E|$ elements and rank $|V|-1$. Every graphic (oriented) matroid can be represented by a connected (directed) graph.

It is known that the cycle oriented matroid of the complete bipartite graph $K_{n,r+1}$ directed from one part to the other equals the oriented matroid of affine dependences between vertices of the product of two simplices of dimensions $n-1$ and $r-2$. This follows, for example, from the description of the product of two simplices which appears in [**19**, pages 246–251].

The remarks before the statement imply that whenever $\mathcal{M}$ is graphic, $\Lambda(\mathcal{M})$ is the cycle oriented matroid corresponding to a connected restriction of the graph $K_{n,r+1}$ which uses all the vertices of the graph. That is, $\Lambda(\mathcal{M})$ is a full-rank restriction of the cycle oriented matroid of $K_{n,r+1}$. This proves (i)$\Rightarrow$(ii).

The other implication is trivial, since every minor of a graphic oriented matroid is graphic and $\mathcal{M}$ is a contraction of $\Lambda(\mathcal{M})$. □

CHAPTER 5

Lifting Triangulations

5.1. Some properties. Lifting versus regular triangulations.

We start by showing the relation between *regular triangulations* of a point configuration and lifting triangulations of the underlying oriented matroid.

EXAMPLES 5.1. (Regular and lifting triangulations)

Let $\mathcal{M}$ be an oriented matroid realized as a point (or vector) configuration $\mathcal{A}$. We think of $\mathcal{A}$ as being represented by a matrix with n columns and r rows, where n is the number of elements of $\mathcal{M}$ and r its rank. A *Gale transform* $\mathcal{A}^*$ of $\mathcal{A}$ is an $n \times (n-r)$ matrix whose row space is the orthogonal complement of the row space of $\mathcal{A}$. It is well-known that $\mathcal{A}^*$ realizes the dual oriented matroid $\mathcal{M}^*$. Any point lying in the convex hull (more generally, in the positive span) of the columns of $\mathcal{A}^*$ defines an interior and realizable extension of $\mathcal{M}^*$. If the point is not in the convex hull of any non-full-dimensional geometric simplex with vertices in $\mathcal{A}^*$, then the extension defines a lifting triangulation of $\mathcal{M}$. The triangulations obtained in this way are called *regular triangulations* of $\mathcal{A}$ [**24**, Definition 1]. Some authors use the word *coherent* [**19**, Chapter 7]. Observe that they can be alternatively defined as those which agree with the projection of the lower envelope of a certain orthogonal lift of $\mathcal{A}$; the equivalence between the two definitions was proved in [**24**], and it is the "realized analogue" of Proposition 4.2.

Regular triangulations are a class in-between lifting and lexicographic triangulations. Their main draw-back in the context of this paper is that regularity depends on the specific realization and not only on the oriented matroid.

Two points in the convex hull of $\mathcal{A}^*$ define the same regular triangulation of $\mathcal{A}$ if and only if they are contained in the same collection of convex hulls of geometric simplices of $\mathcal{A}^*$; that is, if the two points are in the same full-dimensional cell of the common refinement of all the triangulations of $\mathcal{A}^*$. This common refinement is the *chamber complex* of $\mathcal{A}^*$. The bijective correspondence between regular triangulations of a configuration $\mathcal{A}$ and maximal chambers of of its Gale transform $\mathcal{A}^*$ was explored in [**6**] and generalized in [**14**] to include a correspondence between non-regular triangulations of $\mathcal{A}$ and "virtual" chambers of $\mathcal{A}^*$, by means of the realized version of Theorem 3.8.

The relation between regular and lifting triangulations divides the triangulations of a realizable oriented matroid $\mathcal{M}$ in four categories, with different degrees of "realizability". First, there are the triangulations which are regular for any realization of $\mathcal{M}$; this includes lexicographic triangulations but also some non-lexicographic ones, as the one shown in part (c) of Figure 5.1. Second, there are triangulations which are regular or non-regular depending on the realization of $\mathcal{M}$, as the ones in parts (a) and (b), in which the oriented matroid is the same but only

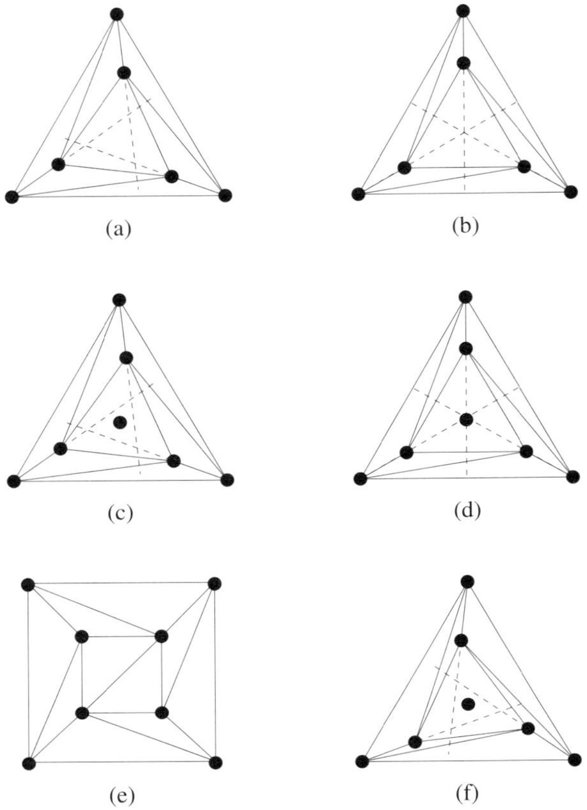

FIGURE 5.1. Some lifting triangulations.

the triangulation in part (a) is regular. These triangulations correspond to extensions of the dual oriented matroid $\mathcal{M}^*$ which are realizable but not as an extension of an arbitrary realization of $\mathcal{M}^*$.

Finally, the triangulations which are not regular for any realization of $\mathcal{M}$, as the ones in parts (d), (e) and (f) of Figure 5.1, can still be lifting triangulations (corresponding to non-realizable lifts/extensions) or not. The three in the figure are lifting triangulations, as follows from the following consequence of Corollary 7.3.2 of [**11**]: if the hyperplanes (flats of corank 1) of an oriented matroid $\mathcal{M}$ which are dependent are all circuits then an arbitrary perturbation of them into bases is an oriented matroid. Applying the corollary to the lift of (d) into a triangular prism with an interior point we get lifts for the triangulations (d) and (f). For (e) we do the same with the lift into a cube. In Examples 5.5 we will construct two very simple non-lifting triangulations and in Section 5.2 we will show some more complicated ones.

We address now the following problem: suppose that we are given an oriented matroid $\mathcal{M}$ of rank r, an element a of $\mathcal{M}$ and a triangulation of either $\mathcal{M}/a$ or $\mathcal{M} \setminus a$; we want to extend it to $\mathcal{M}$. More precisely, for a triangulation T' of $\mathcal{M} \setminus a$

we want to find a triangulation T of $\mathcal{M}$ with $T' \subseteq T$. For a triangulation T'' of $\mathcal{M}/a$ we want to find a triangulation T of $\mathcal{M}$ with $T'' = link_T(a)$. We also want to know if good lifting properties of T' and T'' can be inherited by T.

Proposition 2.10 was a first result in this direction: a triangulation of $\mathcal{M} \setminus a$ can always be extended to $\mathcal{M}$. However, the property fails in general for triangulations of $\mathcal{M}/a$ as the following example of an oriented matroid of rank 4 in 7 elements shows: Consider the point configuration in part (f) of Figure 5.1, and "lift it" to a point configuration in $\mathbb{R}^3$ by giving three different heights to the seven points; put the three vertices of the outer triangle on the bottom, the interior point on top, and the three vertices of the inner triangle in the middle, very close to the bottom. Let a be the point on top and consider the seven tetrahedra obtained by coning a to the seven triangles which appear in the figure. This collection of tetrahedra is known not to be completable to a triangulation of the point configuration. Observe that the link of the top point in the non-completable collection of simplices is a lifting (but non-regular) triangulation of the vertex figure. In fact, it is the same triangulation of parts (a) and (b) of Figure 5.1, in a different realization of the oriented matroid.

This same example has appeared in the proof of Lemma 2.1 in [**14**], and different versions of it have appeared in other places, going back to Schönhardt [**37**]. The mentioned Lemma 2.1 of [**14**] also says that *regular* triangulations of both a deletion and a contraction of $\mathcal{M}$ can be extended to regular triangulations of $\mathcal{M}$. Since lifting triangulations are in some sense the oriented matroid analogue of regular triangulations (see Examples 5.1), one could expect that the same holds for lifting triangulations. The example above shows that this is not the case and the following proposition tells us what is true.

PROPOSITION 5.2. *Let $\mathcal{M}$ be an oriented matroid of rank r on a set E and let $a \in E$ be one of its elements.*

(i) *Let T' be a lifting triangulation of the contraction $\mathcal{M}/a$. Suppose that either T' is a lexicographic triangulation or $\mathcal{M}^*$ is a lexicographic extension of $\mathcal{M}^* \setminus a$. Then, there is a lifting triangulation T of $\mathcal{M}$ such that for every simplex τ in T' the simplex $\{a\} \cup \tau$ is in T. Moreover, if T' is lexicographic, then T can also be taken lexicographic.*

(ii) *Let T' be a lifting triangulation of the deletion $\mathcal{M} \setminus a = \mathcal{M}(E \setminus a)$. Then, there is a lifting triangulation T of $\mathcal{M}$ with $T' \subseteq T$. Moreover, if T' is lexicographic, then T can also be taken lexicographic.*

PROOF. (i) The lifting triangulation T' of $\mathcal{M}/a$ corresponds to an extension $(\mathcal{M}^* \setminus a) \cup p$ of $\mathcal{M}^*/a$. Thus, we have two extensions of $\mathcal{M}^* \setminus a$ (by the elements p and a). It suffices to show that they are "compatible"; that is, that there is a two-element extension $(\mathcal{M}^* \setminus a) \cup \{p, a\}$ whose deletions by p and a coincide respectively with $\mathcal{M}^*$ and $(\mathcal{M}^* \setminus a) \cup p$. This is true if one of the extensions is lexicographic, which is our hypothesis, by Lemma 1.9.

(ii) Let $\mathcal{M}^*$ be the dual oriented matroid to $\mathcal{M}$. Then $\mathcal{M}^*/a$ is the dual of the deletion $\mathcal{M} \setminus a$. Let $(\mathcal{M}^*/a) \cup p$ be the interior extension in general position which defines the triangulation T'. We need to find an interior extension $\mathcal{M}^* \cup p'$ in general position of $\mathcal{M}^*$ such that whenever τ is a maximal simplex of $\mathcal{M}^*/a$ having p in its convex hull, $\tau \cup \{a\}$ is a maximal simplex of $\mathcal{M}^*$ having p' in its convex hull. Such an extension was constructed in Lemma 1.10. □

The following statement can be rephrased saying that a lifting subdivision of $\mathcal{M}$ induces a lifting subdivision of every minor of $\mathcal{M}$. It will be convenient to use the following notations, where S is any collection of subsets of the ground set E of $\mathcal{M}$, τ is a subset of some element of S, and $A \subseteq E$. The last two specialize to the use of link and $\mathcal{P}$ in Definition 2.3 and Section 2.4, respectively, if S is simplicial:

$$S|_A := \{\tau \cap A : \tau \in S\}, \qquad \text{link}_S(\tau) := \{\sigma \setminus \tau : \tau \subset \sigma, \sigma \in S\},$$

$$\mathcal{P}(S) := \{\tau \subseteq E : \tau \text{ is a face of } \mathcal{M}(\sigma) \text{ for some } \sigma \in S\}$$

Observe that $\mathcal{P}(S)|_A \subseteq \mathcal{P}(S|_A)$. The converse holds if S is simplicial, but not in general.

PROPOSITION 5.3. *Let S be a lifting subdivision of an oriented matroid $\mathcal{M}$ on a ground set E, then:*

(i) *For every $A \subset E$ there is a lifting subdivision S_A of $\mathcal{M}(A)$ such that $\mathcal{P}(S)|_A \subseteq \mathcal{P}(S_A)$ (in particular, S_A contains all the cells of S contained in A). If S is a triangulation, S_A can be taken to be a triangulation too.*
(ii) *$link_S(\tau)$ is a lifting subdivision of $\mathcal{M}/\tau$, for every τ contained in some cell of S.*

PROOF. Let $\widehat{\mathcal{M}}$ be a non-cyclic lift of $\mathcal{M}$ on the set $E \cup \{\widehat{p}\}$ which defines the lifting subdivision S of $\mathcal{M}$. If S is a triangulation, we can assume $\widehat{\mathcal{M}}$ to be generic, meaning that its dual is an extension in general position, by part (i) of Proposition 4.2. Then, $\mathcal{P}(S)$ is the collection of faces of $\widehat{\mathcal{M}}$ which do not contain $\widehat{p}$.

(i) For every $A \subseteq E$ the restriction $\widehat{\mathcal{M}}(A \cup \{p\})$ is a non-cyclic lift of $\mathcal{M}(A)$ and defines a subdivision S_A of $\mathcal{M}(A)$. The equation $\mathcal{P}(S)|_A \subseteq \mathcal{P}(S_A)$ is just the fact that every face of $\widehat{\mathcal{M}}$ not containing p intersected with A gives a face of $\widehat{\mathcal{M}}(A \cup \{p\})$. If S was a triangulation and the lift is generic, then S_A is a triangulation.

(ii) Similarly, if τ is contained in a face of of $\widehat{\mathcal{M}}$ not containing p, then $\widehat{\mathcal{M}}/\tau$ is a non-cyclic lift of $\mathcal{M}/\tau$ whose faces are the link of τ in the complex of faces of $\widehat{\mathcal{M}}$. Hence, it produces the subdivision $link_S(\tau)$ of $\mathcal{M}/\tau$. □

It is easy to construct triangulations failing to fulfil the conclusion of part (i) of Proposition 5.3, which hence are non-lifting triangulations. This is done in Examples 5.5. It is an open question whether every non-lifting triangulation can be proved to be non-lifting in this fashion.

Another use of Proposition 5.3 is to characterize lifting subdivisions of $\mathcal{M}$ as the ones which are links of subdivisions of the Lawrence polytope $\Lambda(\mathcal{M})$. It is a bit surprising that the concept of liftingness, heavily based in oriented matroid theory, can be characterized in this simple way and (in the realizable case) with no mention of oriented matroids at all. Another such characterization will be given in Section 5.3.

PROPOSITION 5.4. *Let S be a subdivision of an oriented matroid $\mathcal{M}$ on a set E. Let $\Lambda(\mathcal{M})$ be the Lawrence polytope associated with $\mathcal{M}$. Recall that $\Lambda(\mathcal{M})$ has element set $E \cup \overline{E}$. The following conditions are equivalent:*

(a) *S is a lifting subdivision.*
(b) *There is a subdivision $\Lambda(S)$ of $\Lambda(\mathcal{M})$ such that $\overline{E}$ is contained in some cell of $\Lambda(S)$ and with $link_{\Lambda(S)}(\overline{E}) = S$. If S is a triangulation then $\Lambda(S)$ can be taken to be a triangulation.*

PROOF. (b)⇒(a) $\Lambda(S)$ is a lifting subdivision (Theorem 4.14) and Proposition 5.3(i) says that every link in a lifting subdivision is a lifting subdivision.

(a)⇒(b) Let $\widehat{\mathcal{M}}$ be a non-cyclic lift producing the lifting subdivision S, and assume that it is generic if S is a triangulation. We call $\widehat{\Lambda(\mathcal{M})}$ the lift of $\Lambda(\mathcal{M})$ dual to the extension $\widehat{\mathcal{M}}^*$ of $\mathcal{M}^*$ (we are using the canonical identification between extensions of $\mathcal{M}^*$ and of $\Lambda(\mathcal{M})^*$).

Since $\widehat{\mathcal{M}}$ is a non-cyclic lift, there is a positive circuit in $\widehat{\mathcal{M}}^*$ containing the new element p. This is still a positive circuit of $\widehat{\Lambda(\mathcal{M})}^*$. Hence, $\widehat{\Lambda(\mathcal{M})}$ has a face containing $\overline{E}$ and not containing $\widehat{p}$. This proves that $\widehat{\Lambda(\mathcal{M})}$ is a non-cyclic lift and that its associated lifting subdivision has a cell containing $\overline{E}$. This lift restricts to $\widehat{\mathcal{M}}$ and, with the same arguments as in the proof of Proposition 5.3, the lifting subdivision $\Lambda(S)$ of $\Lambda(\mathcal{M})$ produced by it has $S = \mathrm{link}_{\Lambda(S)}(\overline{E})$ □

EXAMPLES 5.5. (Two non-lifting triangulations)

Let $\mathcal{A}$ be the point configuration in $\mathbb{R}^3$ obtained by giving three different heights to the seven points of Figure 5.1(d): the inner triangle at the bottom, the outer triangle in the middle and the middle point on top. Call $\mathcal{M}$ the rank 4 oriented matroid with 7 elements obtained. Let T be the triangulation of $\mathcal{A}$ (and of $\mathcal{M}$) obtained coning each triangle of the planar triangulation in the figure to the top point. Proposition 5.3 implies that T is not lifting, since removing the top point we get a triangulation of a part of the boundary of a triangular prism which cannot be extended to the whole prism.

Another proof of the fact that T is non-lifting is as follows: in any realization of $\mathcal{M}$, the geometric link in T of the top point is precisely the triangulation of part (b) of Figure 5.1, with the three dashed lines converging in one point thanks to Desargues theorem. Thus, this link is non-regular. A triangulation with a non-regular link is itself non-regular, by (the proof of) Lemma 2.1 of [**14**]. Thus, T is non-regular in every realization of $\mathcal{M}$. Since $\mathcal{M}$ has rank 4 and 7 elements, every extension of its dual is realizable. Thus, every lifting triangulation of $\mathcal{M}$ is regular for some realization of $\mathcal{M}$, which proves that T is not a lifting triangulation.

The same construction of a non-lifting triangulation applies to part (f) of the figure, except that now three additional simplices have to be added in order to fill completely the convex hull of the lifted point configuration. The non-lifting triangulation obtained in this case has the strong property that it is non-lifting for any oriented matroid of which it is a triangulation (the other one does not, as follows from the same construction applied to the regular triangulation (c), which produces a combinatorially equivalent regular triangulation). The proof of this fact is as follows: let $\mathcal{M}$ be an oriented matroid having this triangulation, and extend it with a new element p opposite to any relative interior extension of $\mathcal{M}$. The whole boundary of $\mathcal{M}$ is visible from p, with the meaning of Proposition 2.10. The triangulation obtained by coning p to the boundary of our triangulation is combinatorially the Barnette 3-dimensional sphere. It is proved in [**11**, Proposition 9.5.3] that this sphere is not the face lattice of any rank 5 matroid polytope. Hence the triangulation on $\mathcal{M} \cup p$ is not lifting. Since it is the only one which extends our triangulation (Proposition 2.10), our triangulation is not lifting (Proposition 5.3(i)). A triangulation of a matroid polytope with this same property (and with almost the same proof) but with one more vertex and seven more maximal simplices appears in Proposition 9.6.4 of [**11**].

5.2. Three interesting non-lifting triangulations

5.2.1. The Edmonds-Fukuda-Mandel oriented matroid. We consider here the oriented matroid EFM(8) which appeared in [**17, 18**]. A detailed study of it can be found also in pages 461–468 of [**11**]. With this oriented matroid we will show strange behaviour of triangulations in the following two respects:

- Suppose that two full-rank disjoint simplices σ_1 and σ_2 of an oriented matroid $\mathcal{M}$ are "strongly separated"; by this we mean that there is a covector which is positive in one and negative in the other. If the dual oriented matroid $\mathcal{M}^*$ is realizable (more generally, if it has the *generalized Euclidean intersection property* $\mathtt{IP}_2$, see Definition 1.11) then there must be a lifting triangulation of $\mathcal{M}$ containing the simplices σ_1 and σ_2. The proof of this is as follows: there is no loss of generality in assuming that $E = \sigma_1 \cup \sigma_2$, since the procedure described in Proposition 2.10 extends lifting triangulations to lifting triangulations. Hence, our hypothesis is that (σ_1, σ_2) is a vector in $\mathcal{M}^*$. Since $\mathcal{M}^*$ has the property $\mathtt{IP}_2$, there is an extension of $\mathcal{M}^*$ which lies in the relative interior of both σ_1 and σ_2 (this is a generalization of Lemma 1.12 whose prove we omit). Thus, there is a lifting triangulation of $\mathcal{M}$ containing the simplices σ_1 and σ_2.

  In EFM(8) there are two strongly separated simplices which do not appear simultaneously in any lifting triangulation. (part (i) of Proposition 5.6).
- In oriented matroids satisfying $\mathtt{IP}_2$, saying that two simplices σ_1 and σ_2 intersect properly is equivalent to non-existence of a circuit with its positive part contained in σ_1 and its negative part contained in σ_2, again by Lemma 1.12.

  This is not true if one does not have $\mathtt{IP}_2$: in EFM$(8)^*$ there are two simplices σ_1 and σ_2 which intersect properly (in fact, there is no extension in the convex hull of both) but (σ_1, σ_2) is a vector (part (ii) of Proposition 5.6).

These two situations are more or less dual to one another: if two full-rank simplices σ_1 and σ_2 of a uniform oriented matroid $\mathcal{M}$ contain the positive and negative part of a circuit, their complements are two (weakly) separated simplices in the dual oriented matroid $\mathcal{M}^*$. One would expect that σ_1^* and σ_2^* lie in some triangulation of $\mathcal{M}^*$, even if it is not a lifting one. If this is the case, Theorem 2.4(g) implies that σ_1 and σ_2 do not lie in a triangulation of $\mathcal{M}$. This happens in our example.

EFM(8) is a rank 4 non-realizable oriented matroid on eight elements $\{1, 2, 3, 4, 5, 6, f, g\}$. As a way of definition, we show in Figure 5.2 the contractions of EFM(8) at the six first elements; these contractions are acyclic oriented matroids of rank 3 on seven elements, realizable as point configurations in the plane. The cocircuits of EFM(8) can be read from the figure: the contraction EFM$(8)/a$ permits to read the cocircuits which do not have a on the support, and no cocircuit can have all the six elements 1, 2, 3, 4, 5, and 6 on its support. The interested reader can check that the cocircuits read from the figure coincide with the ones listed in page 464 of [**11**].

The figure shows that EFM(8) has symmetry group isomorphic to S_3, generated by the permutations $(16)(24)(35)(fg)$ and $(123)(456)$.

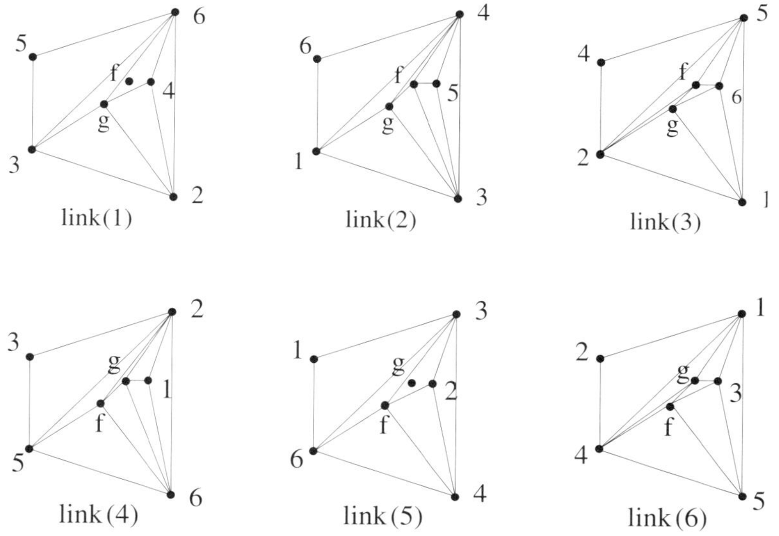

FIGURE 5.2. Links of a triangulation of EMF(8).

In the figure we have drawn a certain triangulation of each contraction. The six triangulations are the links at the vertices $1, \ldots, 6$ of the following collection of simplices of EFM(8):

$$T := \{\{1624\}, \{2435\}, \{3516\},$$
$$\{235f\}, \{245f\}, \{356f\}, \{456f\},$$
$$\{146g\}, \{136g\}, \{124g\}, \{123g\},$$
$$\{24fg\}, \{46fg\}, \{63fg\}, \{32fg\}\}.$$

Claim 1: T is a triangulation of EFM(8).

PROOF. Any interior corank 1 simplex τ contains at least one element $a \in \{1, 2, 3, 4, 5, 6\}$. The oriented pseudo-manifold property for $\tau \setminus \{a\}$ in $link_T(a)$ implies the property for τ in T. Thus, T has the oriented pseudo-manifold property. Also, since 1 is a vertex of EFM(8), any simplex of T which covers a lexicographic interior extension in general position starting by $[1^+, \ldots]$ must contain 1. The fact that $link_T(1)$ is a triangulation implies that all such extensions are covered exactly once. □

Claim 2: T is a non-lifting triangulation of EFM(8).

PROOF. We look at the restriction of EFM(8) to the elements $\{1,2,3,4,5,6\}$. The restriction of T contains the simplices $\{1624\}$, $\{2435\}$ and $\{3516\}$. Suppose that there is a triangulation T' of the restriction containing these three simplices. For the link of T' at 1 to be a triangulation it is necessary that $\{1346\}$ and $\{1234\}$ be in T'. But in the contraction at element 3 we see that $\{1234\}$ intersects $\{2345\}$ improperly.

That is, the restriction of T to the six elements cannot be extended to a triangulation. Proposition 5.3 implies that T is not a lifting triangulation. □

Claim 3: *T is the only triangulation of* EFM(8) *containing the simplices $\{146g\}$ and $\{235f\}$.*

PROOF. We will show that the only way to complete $\{146g\}$ and $\{235f\}$ to a triangulation of EFM(8) is using precisely the simplices of T.

- the presence of $\{146g\}$ implies (see the contraction at element 1) the presence of the simplex $\{1246\}$ and the absence of any simplex containing $\{1f\}$. With similar arguments at 5 we conclude the presence of $\{2345\}$ and the absence of $\{5g\}$.
- then, the fact that 5 and g do not lie on the same simplex in the contraction at element 1 implies the presence of the simplices $\{136g\}$ and $\{1356\}$. A similar argument at 5 shows the presence of $\{356f\}$.
- with this, the only way to complete the links at 3 and 6 to subdivisions is the inclusion of the simplices $\{36fg\}$, $\{23fg\}$, $\{123g\}$, $\{46fg\}$ and $\{456f\}$.
- the simplex $\{124g\}$ completes the link at 1, the simplex $\{245f\}$ completes the link at 5 and then $\{24fg\}$ completes the links at 2 and 4. □

The three claims together imply that no lifting triangulation of EFM(8) contains the two simplices $\{146g\}$ and $\{235f\}$. Part (ii) of the following statement has the following stronger implication: no lifting subdivision of EFM(8) has $\{146g\}$ and $\{235f\}$ contained in two different cells.

PROPOSITION 5.6. (i) *No lifting triangulation of* EFM(8) *contains the two simplices $\{146g\}$ and $\{235f\}$, although $(\{146g\},\{235f\})$ is a covector.*

(ii) *The dual oriented matroid* $\mathrm{EFM}(8)^*$ *has no extension* $\mathrm{EFM}(8)^*\cup p$ *with* $p \in conv_{\mathrm{EFM}(8)^*\cup p}(\{146g\}) \cap conv_{\mathrm{EFM}(8)^*}(\{235f\})$. *In particular, the two simplices $\{146g\}$ and $\{235f\}$ intersect properly in* $\mathrm{EFM}(8)^*$, *although $(\{16g\},\{35\})$ is a circuit. No triangulation of* $\mathrm{EFM}(8)^*$ *contains both simplices.*

PROOF. (i) The covector $(\{16g\},\{235f\})$ can be read from the contraction at element 4 in Figure 5.2. Its composition with any covector positive at 4 shows that $(\{146g\},\{235f\})$ is a covector. The rest of the statement follows from claims 2 and 3.

(ii) That $(\{16g\},\{35\})$ is a cocircuit of EFM(8) can be read from the contraction at either 2 or 4. There is no triangulation of $\mathrm{EFM}(8)^*$ containing $\{146g\}$ and $\{235f\}$ because such a triangulation would contain the complements of two simplices of the triangulation T of EFM(8), which is impossible by part (g) of Theorem 2.4. We now prove that the two simplices intersect properly; even more, that there is no extension $\mathrm{EFM}(8)^*\cup p$ with $p \in conv(\{146g\}) \cap conv(\{235f\})$.

If there was one such extension $\mathrm{EFM}(8)^*\cup p$, then there would be two circuits $(\tau_1,\{p\})$ and $(\tau_2,\{p\})$ in $\mathrm{EFM}(8)^*\cup p$ with $\tau_1 \subseteq \{146g\}$ and $\tau_2 \subseteq \{235f\}$. Elimination of the element p would imply that (τ_1, τ_2) is a vector of $\mathtt{EFM}(8)^* \cup p$ and hence also of $\mathtt{EFM}(8)^*$. Since $\mathtt{EFM}(8)^*$ is uniform of rank four, $\tau_1 \cup \tau_2$ has at least five elements and one of τ_1 or τ_2 (say τ_1) has at least three elements.

If τ_1 has four elements, then p is in the relative interior of $\{146g\}$. The perturbation $p' := [p^+, 2^+, 3^+, 5^+, f^+]$ will produce an extension of $\mathtt{EFM}(8)^*$ interior, in general position and in the relative interior of both $\{146g\}$ and $\{235f\}$. This is impossible since then the associated lifting triangulation of $\mathtt{EFM}(8)$ contains both simplices, in contradiction with claims 1 and 2.

If τ_1 has three elements, then τ_2 is a positive vector in the rank-1 contraction $\mathtt{EFM}(8)^*/\tau_1$. In particular, there is an element $a \in \tau_2$ such that the cocircuits vanishing on τ_1 have the same sign on a and in the element $\{146g\} \setminus \tau_1$. Then, the perturbation $p' := [p^+, a^+]$ is still in the convex hull of τ_2 and in the relative interior of $\{146g\}$. This is the previous case. □

Another feature of the example $\mathtt{EFM}(8)$ is the following. Its deletion at element f is realizable by the columns of the following matrix, as shown in page 461 of [**11**]:

$$\begin{pmatrix} -1 & \epsilon & -\epsilon & 1 & 0 & 0 & 0 \\ -\epsilon & -1 & \epsilon & 0 & 1 & 0 & 0 \\ \epsilon & -\epsilon & -1 & 0 & 0 & 1 & 0 \\ 1 & 1 & 1 & 0 & 0 & 0 & 1 \end{pmatrix}.$$

A Gale transform of this configuration, rescaled at the element g (in particular, reoriented) is given by:

$$\begin{pmatrix} 1 & 0 & 0 & 1 & \epsilon & -\epsilon & 1/3 \\ 0 & 1 & 0 & -\epsilon & \epsilon & 1 & 1/3 \\ 0 & 0 & 1 & \epsilon & 1 & -\epsilon & 1/3 \end{pmatrix},$$

which can be viewed as a point configuration $\mathcal{A}$ in the plane $x + y + z = 1$ of $\mathbb{R}^3$. $\mathcal{A}$ consists of the six vertices of a (non-regular) hexagon and an interior point in it. Using the pushing-pulling characterization of lexicographic triangulations (see Remark 4.4) it is easy to conclude that all the triangulations of $\mathcal{A}$ are lexicographic (and thus regular). However, the dual oriented matroid has non-realizable extensions (such as the reorientation at g of $\mathtt{EFM}(8)$). The apparent contradiction is not so, because different extensions of the dual can correspond to the same triangulation of $\mathcal{A}$.

5.2.2. A non-lifting triangulation of a unimodular polytope. We call a real matrix *unimodular* if all its maximal non-zero minors have the same absolute value (sometimes the definition asks this value to be 1). It is a classical result of matroid theory that a matroid can be represented over the rationals by a unimodular matrix if and only if it can be represented over any field [**27**, Theorem 6.6.3]; such matroids are called *regular*. For orientable matroids this is equivalent to the matroid being *binary*, i.e. representable over the field with two elements [**27**, Corollary 13.4.6], and also equivalent to the absence of any rank 2 uniform minor on 4 elements [**27**, Theorem 6.5.4].

A vector configuration is unimodular if it equals the set of columns of a unimodular matrix. A point configuration of dimension d is unimodular if all its full-dimensional simplices have the same volume. Equivalently, if adding an extra constant coordinate to every point one gets a unimodular vector configuration. Unimodular vector or point configurations play an important role in different branches of discrete mathematics. In connection with triangulations, it was somewhat unexpected that unimodular configurations can have non-regular triangulations. This was shown by de Loera [**13**] who constructed a non-regular triangulation of the product $\Delta_3 \times \Delta_3$ of two 3-dimensional simplices. Later on Sturmfels [**38**, Theorem 10.15] constructed a non-regular triangulation of $\Delta_2 \times \Delta_5$. Eric Babson (personal communication) has shown that de Loera's triangulation of $\Delta_3 \times \Delta_3$ is, in fact, a non-lifting triangulation. Here we will construct a non-lifting triangulation of a different unimodular polytope.

Before going into detail let us make another consideration. Let $v_1, \dots, v_n \in \mathbb{R}^r$ be the columns of a rank r acyclic unimodular matrix. Let K be the common absolute value of all its non-zero maximal minors. By acyclic we mean that there is a linear functional on $\mathbb{R}^r$ positive in all the columns of the matrix. Let f be such a functional. Dividing each vector v_i by the value $f(v_i)$ we get a homogeneous matrix and thus a point configuration in $\mathbb{R}^{r-1}$, identifying $\mathbb{R}^{r-1}$ with the affine hyperplane $\{f(v) = 1)\}$ in $\mathbb{R}^r$. We say that this configuration is *weighted unimodular* with weights $f(v_1), \dots, f(v_n)$ because the volume of any maximal simplex of the configuration multiplied by the product of weights of its vertices equals $K/d!$. If all the weights are rational (which can be achieved by a rational choice of f if $\mathcal{A}$ itself is rational), the lifting procedure exhibited in Section 4.4 allows to construct a truly unimodular configuration with "the same" triangulations as $\mathcal{A}$. This follows from the following result:

PROPOSITION 5.7. *Let $\{P_1, \dots, P_n\}$ be a weighted unimodular point configuration in $\mathbb{R}^{r-1}$ with positive integer weights $w_1, \dots, w_n$. Let $\mathcal{M}$ be the oriented matroid of affine dependences of the point configuration and consider the oriented matroid $\Sigma(\mathcal{M})$ of Section 4.4 taking each k_i to be precisely the weight w_i. Then, $\Sigma(\mathcal{M})$ is the oriented matroid of affine dependences of a certain unimodular point configuration with $\sum w_i$ points in dimension $\sum w_i - n + r - 1$.*

PROOF. We will explicitly construct the matrix of a certain point configuration of the stated rank and number of columns, and then show that it is unimodular, homogeneous and that it realizes $\Sigma(\mathcal{M})$.

Let $v_1, \dots, v_n$ be a non-homogeneous unimodular vector configuration in $\mathbb{R}^r$ which homogenizes to $P_1, \dots, P_n$. Let $\{v_1^*, \dots, v_n^*\} \in \mathbb{R}^{n-r}$ be a Gale transform of it. That is, assume that the matrices A and A' having as columns the vectors v_i and v_i' respectively have orthogonally complementary row spaces.

For each $i = 1, \dots, n$ we consider the following three matrices, of sizes $(\omega_i - 1) \times w_i$, $r \times w_i$ and $(n-r) \times w_i$ respectively. $I_{\omega_i - 1}$ represents the identity matrix of size $\omega_i - 1$ and $1_{\omega_i - 1}$ the column vector $(1, 1, \dots, 1)$ of length $w_i - 1$:

$$S_i := (-I_{\omega_i - 1} : 1_{\omega_i - 1}) \qquad V_i := (v_i/w_i \ \cdots \ v_i/w_i) \qquad V_i^* := (v_i^* \ \cdots \ v_i^*).$$

Consider then the following two matrices:

$$\Sigma(A) := \begin{pmatrix} S_1 & 0 & \cdots & 0 \\ 0 & S_2 & \cdots & 0 \\ \vdots & \vdots & \ddots & \vdots \\ 0 & 0 & \cdots & S_n \\ V_1 & V_2 & \cdots & V_n \end{pmatrix} \qquad \Sigma(A)^* := (\, V_1^* \quad \cdots \quad V_n^* \,).$$

It is obvious that $\Sigma(A)^*$ realizes the oriented matroid $\Sigma(\mathcal{M})^*$. It is also easy to check that $\Sigma(A)$ and $\Sigma(A)^*$ are Gale transforms of one another: the lower rows of

$\Sigma(A)$ are orthogonal to the rows of $\Sigma(A)^*$ because A and A^* are Gale transforms of one another, while the upper rows of $\Sigma(A)$ are orthogonal to the rows of $\Sigma(A)^*$ because the sum of the entries in each row of each S_i is zero while the rows of each V_i^* are constant. Thus, $\Sigma(A)$ realizes the oriented matroid $\Sigma(\mathcal{M})$.

Let $f = (f_1, \dots, f_r)$ be a linear functional on $\mathbb{R}^r$ which gives the weights $w_i = f(v_i)$. Then, the linear functional $(0, \dots, 0, f_1, \dots, f_r)$ with a string of $\sum w_i - n$ zeroes shows that the columns of $\Sigma(A)$ are homogeneous.

Unimodularity of $\Sigma(A)$ can be checked directly, or deduced from the fact that a Gale transform of a unimodular matrix is unimodular as well. This implies that A^* is unimodular, which clearly shows unimodularity for $\Sigma(A)^*$ and in turn for $\Sigma(A)$. □

We now construct the desired non-lifting triangulation of a homogeneous polytope. As a first step we construct a non-lifting triangulation of a weighted unimodular point configuration $\mathcal{A}$ with 9 points in $\mathbb{R}^3$. The configuration in question is given by the columns of the following homogeneous rank 4 matrix:

$$A := \begin{pmatrix} 1 & 0 & 0 & 0 & 1/2 & 0 & 0 & 1/2 & 1/4 \\ 0 & 1 & 0 & 0 & 1/2 & 1/2 & 0 & 0 & 1/4 \\ 0 & 0 & 1 & 0 & 0 & 1/2 & 1/2 & 0 & 1/4 \\ 0 & 0 & 0 & 1 & 0 & 0 & 1/2 & 1/2 & 1/4 \end{pmatrix}.$$

Geometrically, the configuration $\mathcal{A}$ consists of the four vertices of a tetrahedron Δ, four mid-points of edges of Δ and the barycenter of Δ. It is weighted unimodular with weights $(1, 1, 1, 1, 2, 2, 2, 2, 4)$, as can be easily checked. The fact that the oriented matroid admits a unimodular representation also follows from the fact that its dual is graphic, and hence binary. Indeed, the dual is the graphic oriented matroid associated to the complete bipartite graph $K_{3,3}$ with the orientation shown in Figure 5.3. The labels on the edges refer to the order of the columns of A.

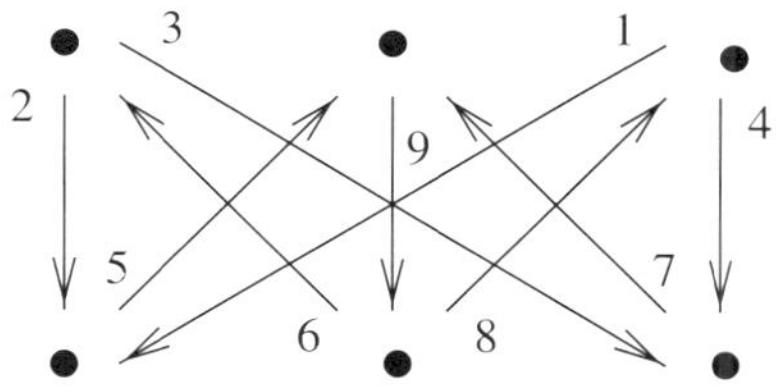

FIGURE 5.3. A graph whose cycle matroid is dual to the configuration $\mathcal{A}$.

Let us consider the following triangulation ∂T of the boundary of the tetrahedron:

$$\partial T := \{\{3,6,7\}, \{2,6,7\}, \{2,4,7\}, \{2,5,6\}, \{1,5,6\}, \{1,3,6\}, \\ \{1,5,8\}, \{4,5,8\}, \{2,4,5\}, \{4,7,8\}, \{3,7,8\}, \{1,3,8\}\}.$$

This triangulation is displayed in Figure 5.4, where the boundary of the tetrahedron appears "unfolded". ∂T cannot be completed to a triangulation T of $\mathcal{A}$ without using the interior point 9: if it could, the triangle $\sigma = \{1,3,6\} \in \partial T$ should be joined in $\widehat{T}$ to one of the three points 4, 7 or 8 which do not lie on the plane containing σ. It can be joined to neither 4 nor 7 because the edges $\{1,4\}$ and

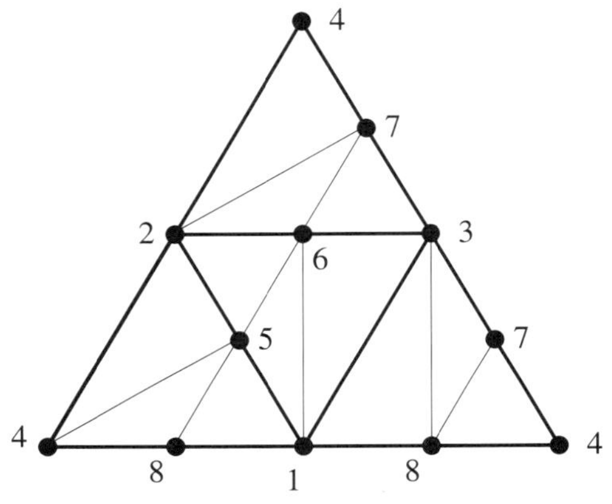

FIGURE 5.4. A triangulation of the boundary of a tetrahedron.

$\{1,7\}$ are not edges of ∂T. Thus, we conclude that $\{6,8\}$ should be an edge of T. With the same arguments we conclude that $\{7,5\}$ should also be an edge, but this is impossible because these two edges intersect improperly.

By Proposition 5.3, this implies that the triangulation T of $\mathcal{A}$ obtained coning ∂T to the central point 9 is not a lifting triangulation of $\mathcal{A}$. Thus:

PROPOSITION 5.8. *$\mathcal{A}$ is a weighted unimodular configuration with 9 points in $\mathbb{R}^3$ which has a non-lifting triangulation. There exists a unimodular polytope with 16 vertices in $\mathbb{R}^{10}$ which has a non-lifting triangulation.*

PROOF. The first part has already been shown. For the second part we apply Proposition 5.7 to $\mathcal{A}$ in order to get a unimodular point configuration with 16 points in $\mathbb{R}^{10}$. The configuration is polytopal by Theorem 4.18, since all the non-vertices of $\mathcal{A}$ have weight at least 2. Also by Theorem 4.18, the non-lifting triangulation described before gives a non-lifting triangulation of the new oriented matroid. □

Incidentally, T is a triangulation with only 4 bistellar flips, supported on the 4 quadrilaterals that appear in the facets of the tetrahedron; that is, the number of flips is less than the dimension $n - d - 1 = 5$ of the associated *secondary polytope* (see the definition in [**6**] or [**19**]). A combinatorially equivalent triangulation with the same property was constructed by de Loera et al. [**15**].

5.2.3. A non-lifting triangulation of the 4-cube. Jesús de Loera [**13**] has shown that the 4-dimensional cube has non-regular triangulations. We go further and construct a non-lifting triangulation of the 4-cube.

We consider the 4-cube as realized by the point configuration C_4 in $\mathbb{R}^4$ whose 16 points are all the 0-1 vectors on 4 coordinates. The contraction of C_4 at any of its points is realized by the point configuration in $\mathbb{R}^3$ consisting of the barycenters of the fifteen faces of a tetrahedron, including the vertices and the tetrahedron itself as faces but excluding the empty face. We observe that the nine-point configuration $\mathcal{A}$ of the previous example is a subconfiguration of this. This fact will be used.

We are first going to show a triangulation of the contraction of C_4 at the point (1111). The points (0011), (1001), (1100) and (0110) are mid-points of edges of the tetrahedron in the contraction, forming a square. The points (1000), (0100), (0010) and (0001) are the barycenters of the four facets of the tetrahedron. These eight points are the vertices of a 3-polytope P with four quadrilaterals and four triangles as facets, depicted in Figure 5.5. We triangulate the configuration formed by these eight points and the central one (which corresponds to (0000)) in the following way: we choose one of the two diagonals in the quadrilateral facets in such a way that each of the vertices of P belongs to exactly one diagonal (which can be done in two equivalent ways; see one of them in Figure 5.5). This triangulates the boundary of P with 12 triangles. Then we cone each of the triangles to the central point (0000).

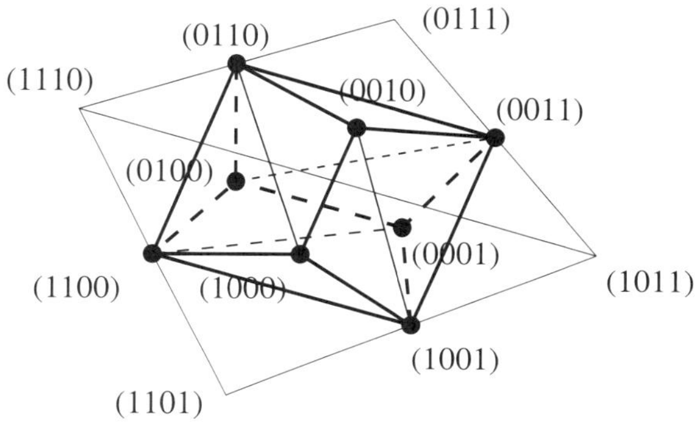

FIGURE 5.5. A triangulation of a polytope P contained in the tetrahedron.

We will complete the triangulation of P to a triangulation of the whole tetrahedron as follows. The mid-point (1010) of the segment [(1011), (1110)] sees two quadrilateral facets of P, and hence four boundary triangles of the triangulation of P. We add the four tetrahedra obtained joining (1010) to them. In the same way, we add the joins of (0101) to the four boundary triangles of the triangulation of P seen from it. More precisely, we are adding the following eight tetrahedra to the triangulation of P:

$$\{\{(1100), (1000), (0110)\}, \{(1000), (0110), (0010)\},$$
$$\{(0011), (0010), (1001)\}, \{(0010), (1001), (1000)\}, \} \cdot (1010)$$

and

$$\{\{(0110), (0100), (0011)\}, \{(0100), (0011), (0001)\},$$
$$\{(1001), (0001), (1100)\}, \{(0001), (1100), (0100)\}, \} \cdot (0101).$$

This produces a triangulation with 20 tetrahedra of the octahedron whose vertices are the mid-points of the edges of the tetrahedron $C_4/(1111)$. It is now easy to complete this to a triangulation of $C_4/(1111)$ with 24 tetrahedra, by adding four tetrahedra. Namely, for each of the vertices (1110), (1101), (1011) and (0111) of the tetrahedron $C_4/(1111)$ we add its cone to the only boundary triangle visible from it.

Thus, we have triangulated $C_4/(1111)$ with 24 tetrahedra. Let us denote by $T(C_4/(1111))$ this triangulation. In C_4, we consider the collection of 24 full-dimensional simplices $T(C_4/(1111)) \cdot (1111)$. It is clear that these 24 simplices intersect properly (in the usual geometric sense) since all of them have (1111) as a vertex and their links at (1111) intersect properly. On the other hand, the volume of every d-simplex in $\mathbb{R}^d$ with integer vertices is an integer multiple of $1/d!$. In particular, the 24 simplices we are considering must cover the 4-cube (again in a geometric sense). We conclude that the 24 simplices provide a triangulation of the 4-cube, which we denote T.

We are now going to prove that T is a non-lifting triangulation. We consider the link of T at (0000), which is a triangulation of the contraction $C_4/(0000)$. In the construction above, the only simplices of T using the point (0000) are the joins to (1111) of the 12 initial tetrahedra used to triangulate the polytope P. Of course, the contraction $C_4/(0000)$ is again the point configuration given by the barycenters of a tetrahedron, but the points which were barycenters of facets in the contraction $C_4/(1111)$ are vertices of the tetrahedron in the contraction $C_4/(0000)$ and vice-versa.

Thus, the 12 simplices of the link at $C_4/(0000)$ are the joins of the central point (1111) with the 12 triangles on the boundary of a tetrahedron which appear in Figure 5.6 (where the boundary of the tetrahedron appears unfolded). But this figure is the same as Figure 5.4. The same argument used there proves that the link of T at (0000) is non-lifting. Thus, T is a non-lifting triangulation of the 4-cube, by Proposition 5.3(ii).

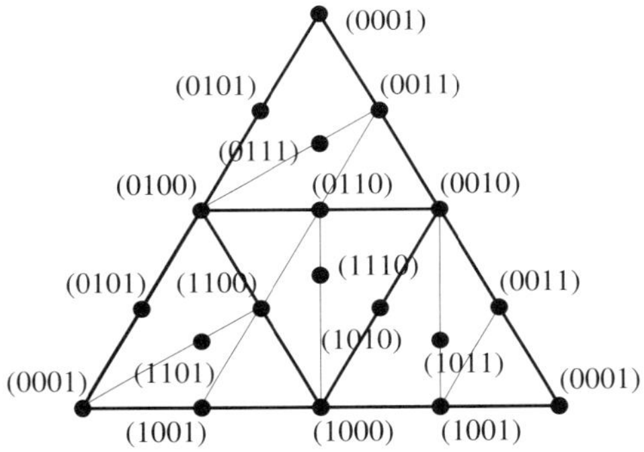

FIGURE 5.6. A triangulation of a link of the 4-cube.

5.3. Two characterizations of lifting subdivisions.

In Section 5.1 we have characterized lifting subdivisions of an oriented matroid $\mathcal{M}$ as those which are links in lifting subdivisions of the Lawrence polytope $\Lambda(\mathcal{M})$ (Proposition 5.4). Here, we give another characterization, related to the fact a lifting subdivision of $\mathcal{M}$ has to be "compatible" with some subdivision of $\mathcal{M}(A)$ for every restriction $\mathcal{M}(A)$ of $\mathcal{M}$ (Proposition 5.3(i)):

DEFINITION 5.9. Let S be a subdivision of an oriented matroid $\mathcal{M}$ of rank r on a ground set E. For each subset $B \subseteq E$, let S_B be a subdivision of $\mathcal{M}(B)$. We say that the family of subdivisions $\mathcal{S} = \{S_B\}_{B \in E}$ is *consistent* if for every $B \subseteq E$ the following happens:

(i) For every cell $\tau \in S_B$ and for every $B' \subset B$, $\tau \cap B'$ is a face of a cell of $S_{B'}$ (i.e., $(S_B)|_{B'} \subset \mathcal{P}(S_{B'})$, with the notation introduced before Proposition 5.3).

(ii) For every rank r simplex σ of $\mathcal{M}(B)$, if σ is contained in a cell of $S_{\sigma \cup \{b\}}$ for every $b \in B \setminus \sigma$, then σ is contained in a cell of S_B as well.

We say that the family *is consistent with* S if, moreover, $S = S_E$.

Condition (i) says that the subcomplex of S_B induced on the elements of any $B' \subset B$ is a subcomplex of $S_{B'}$. Condition (ii) is void unless B spans $\mathcal{M}$ and has at least $r+2$ elements.

THEOREM 5.10. *The following conditions are equivalent for a subdivision S of an oriented matroid $\mathcal{M}$:*

(a) *S is a lifting subdivision.*

(b) *There is a subdivision $\Lambda(S)$ of $\Lambda(\mathcal{M})$ such that $\overline{E}$ is contained in some cell of $\Lambda(S)$ and with $link_{\Lambda(S)}(\overline{E}) = S$. If S is a triangulation then $\Lambda(S)$ can be taken to be a triangulation.*

(c) *There is a family $\mathcal{S}$ of subdivisions of the restrictions of $\mathcal{M}$ which is consistent with S.*

The equivalence of (a) and (b) was proved in Proposition 5.4. For proving the equivalence of (a) and (c) we first show how a consistent family of subdivisions of $\mathcal{M}$ induces a *circuit signature function* λ. I.e. a function assigning a 0, +1 or -1 to each circuit of $\mathcal{M}$ (Lemma 5.11). This will be a cocircuit signature function on $\mathcal{M}^*$, and we prove that it defines an extension of $\mathcal{M}^*$ producing the lifting triangulation we desire. This is done by "reduction to rank 2" (Lemma 5.12).

If B has corank 1, then $\mathcal{M}(B)$ has exactly one circuit $C = (C^+, C^-)$ (up to sign reversal) and three subdivisions, which we denote as follows (if $\mathcal{M}(B)$ is not acyclic then one of $S(B, C^+)$ or $S(B, C^-)$ does not exist, but this does not affect our proof):

$$S(B, C^+) := \{B \setminus \{a\} : a \in C^+\},$$
$$S(B, C^-) := \{B \setminus \{a\} : a \in C^-\}, \qquad S(B, C^0) := \{B\}.$$

We will say that the three subdivisions above give positive, negative and zero sign to the circuit C, respectively.

LEMMA 5.11. *Let S be a subdivision of an oriented matroid $\mathcal{M}$. Let $\mathcal{C}$ denote the set of circuits of $\mathcal{M}$. Let $\mathcal{S} = \{S_B\}_{B \subseteq E}$ be a family of subdivisions of the restrictions of $\mathcal{M}$ which is consistent with S.*

Define a circuit signature function $\lambda_{\mathcal{S}} : \mathcal{C} \to \{-1, 0, +1\}$ as follows. For each circuit C of $\mathcal{M}$, let B be any corank 1 subset of E containing $\overline{C}$. Let $\lambda_{\mathcal{S}}(C)$ be -1, $+1$ or 0 if S_B equals $S(B, C^+)$, $S(B, C^-)$, and $S(B, C^0)$, respectively (observe the sign change). Then,

(i) *The function $\lambda_{\mathcal{S}}$ is well-defined (it does not depend on the choice of the subset B) and satisfies $\lambda_{\mathcal{S}}(-C) = -\lambda_{\mathcal{S}}(C)$.*

(ii) *If $\lambda_{\mathcal{S}}$ defines a lift $\widehat{\mathcal{M}}$ of $\mathcal{M}$ (i.e. if it defines an extension of $\mathcal{M}^*$), then the lift is non-cyclic and S is the lifting subdivision induced by it.*

PROOF. (i) Let C be a circuit of $\mathcal{M}$. Then, $\overline{C}$ is already a corank 1 subset of E containing $\overline{C}$. Moreover, $\overline{C}$ is a face of of B for any other corank 1 subset of E containing $\overline{C}$, because the unique two circuits (C^+, C^-) and (C^-, C^+) of $\mathcal{M}(B)$

are orthogonal to $(\emptyset, B \setminus \overline{C})$. Hence, by condition (i) in Definition 5.9, S_B gives the same sign to the circuit C as $S_{\overline{C}}$, for every such B. That $\lambda(-C) = -\lambda(C)$ is trivial.

(ii) Suppose that $\lambda_{\mathcal{S}}$ defines a lift $\widehat{\mathcal{M}}$ of $\mathcal{M}$. Let $\widehat{p}$ denote the new element. If the lift was cyclic, let $(\emptyset, B \cup \{\widehat{p}\})$ be a positive circuit of it containing $\widehat{p}$. Then, $C = (\emptyset, B)$ would be a positive circuit in $\mathcal{M}$ and we would have $\lambda_{\mathcal{S}}(C) = +1$. This is impossible since the subdivision $S(B, C^-)$ does not exist in these conditions.

Denote $S_{\widehat{\mathcal{M}}}$ the lifting subdivision induced by $\widehat{\mathcal{M}}$.

We first deal with the case of $\widehat{\mathcal{M}}$ being the lift by a coloop. Then, $S_{\widehat{\mathcal{M}}}$ is the trivial subdivision. By property (i) in Definition 5.9 and the definition of $\lambda_{\mathcal{S}}$, we know that every subset of E of corank 1 is contained in some cell of S. We will prove by induction on $|E|$ that S is the trivial subdivision. If $\mathcal{M}$ has corank 0, then this is obvious: its only subdivision is the trivial one. Otherwise, let $a \in E$ be an element such that $\mathcal{M} \setminus a$ has the same rank as $\mathcal{M}$. By inductive hypothesis, $S_{E\setminus\{a\}}$ is the trivial subdivision. Let $\sigma \subseteq E \setminus \{a\}$ be a full-rank simplex. Then, $\sigma \cup \{a\}$ and $E \setminus \{a\}$ are contained in cells of S, the first one because it has corank one and the second by Definition 5.9. But then the definition of subdivision (see part (c) of Definition 4.5) implies that E is (contained in) a cell of S, and hence S is the trivial subdivision.

Now we deal with the general case. Let F be a cell of $S_{\widehat{\mathcal{M}}}$, i.e. let F be a face of $\widehat{M}$ not containing $\widehat{p}$. Then, $\lambda_{\mathcal{S}}$ restricted to F is the zero function, and by the previous case S_F is the trivial subdivision. By Definition 5.9, F is contained in a cell of S, i.e. $S_{\widehat{\mathcal{M}}}$ refines S. We only have to prove that the union of any two different cells of $S_{\widehat{\mathcal{M}}}$ is not contained in S.

Let F_1 and F_2 be two different cells of $S_{\widehat{\mathcal{M}}}$. Let σ be a full rank simplex contained in F_1 and let a be an element in $F_2 \setminus F_1$, so that $\sigma \cup \{a\}$ has corank 1. The subdivision $S_{\sigma\cup\{a\}}$ is the only one having σ as a cell, which implies by Definition 5.9 that the unique cell of S containing F_1 does not contain a. □

LEMMA 5.12. *In the same conditions of Lemma 5.11, suppose moreover that $\mathcal{M}$ has corank 2. Then, the circuit signature $\lambda_{\mathcal{S}}$ induced is the circuit signature of a lift of $\mathcal{M}$.*

PROOF. Without loss of generality, we assume that $\mathcal{M}$ has no coloops. In other words, that for every element $a \in E$ its deletion $\mathcal{M} \setminus a$ has corank 1. Otherwise the statement follows easily by induction on the cardinality of $\mathcal{M}$.

In these conditions, for each element $a \in E$, $\mathcal{M} \setminus a$ has a unique circuit C_a (up to a sign), which is given a certain sign by $\lambda_{\mathcal{S}}$. The dual $\mathcal{M}^*$ of $\mathcal{M}$ has rank 2 and can be realized by a vector configuration of rank 2, whose cocircuits are the complements of the lines generated by vectors of the configuration. We can picture $\lambda_{\mathcal{S}}(C_a)$ by putting a $+$ and a $-$ sign on the two sides of the vector a, in the way indicated by $\lambda_{\mathcal{S}}(C_a)$ if this is non-zero and putting zeroes if $\lambda_{\mathcal{S}}(C_a) = 0$.

We will prove that our circuit signature function does not contain any of the three forbidden subconfigurations displayed in Figure 1.1 (see part (c) of Lemma 1.3). Taking into account the possible reorientations, each of the three subconfigurations breaks into several cases, displayed in Figure 5.7. The pictures in each row are the different reorientations of the same forbidden subconfiguration. We only need to show that none of them can appear in the dual of $\mathcal{M}$, when we picture $\lambda_{\mathcal{S}}$ as indicated above. Observe that a zero in a vector v of the picture means that

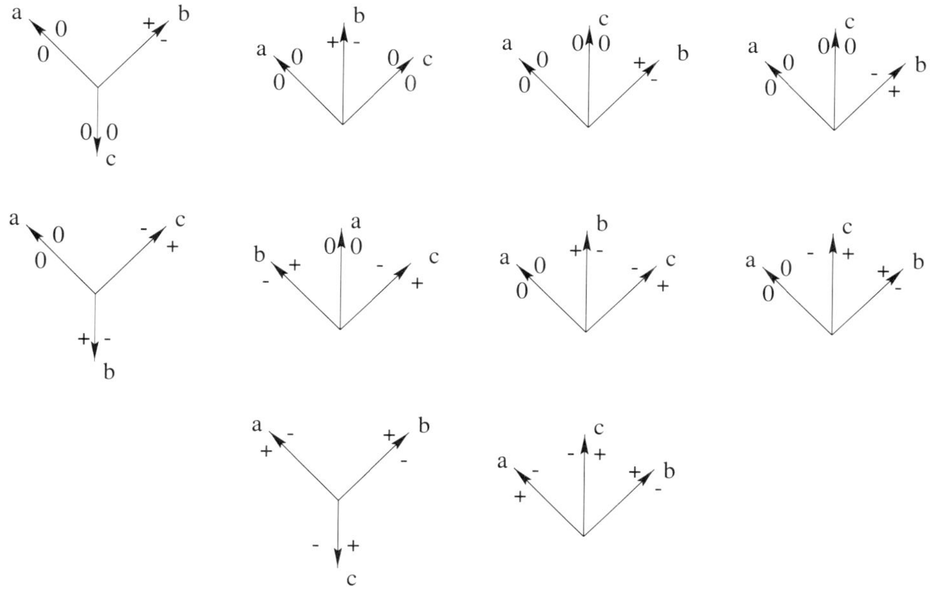

FIGURE 5.7. Forbidden subconfigurations for a cocircuit signature in rank 2.

$S_{\mathcal{M}\setminus\{v\}}$ is a trivial subdivision, while a $+$ on one side of v means that, for every w on the other side of v, $\mathcal{M}\setminus\{v,w\}$ is a cell in $S_{\mathcal{M}\setminus\{v\}}$. With this we can discard the different possibilities as follows:

(1) In the first row of pictures we have zero signs for C_a and C_c, but not for C_b. This implies that $S_{E\setminus\{a\}}$ and $S_{E\setminus\{c\}}$ are trivial subdivisions. Taking $B=E$ in part (ii) of Definition 5.9 we conclude that $E\setminus\{a,c\}$ is contained in a cell σ of S. Taking $B=E$, $B'=E\setminus\{c\}$ and $\tau=\sigma$ in part (i) of Definition 5.9, we have that $a\in\sigma$ (since $S_{B'}$ is the trivial subdivision). In the same way one proves $c\in\sigma$, i.e. $\sigma=E$. But then $S_{E\setminus\{b\}}$ is trivial as well, which is not the case.

(2) In the pictures of the second row we still have that $S_{E\setminus\{a\}}$ is the trivial subdivision. We have labelled all the cases so that the vector a of the Gale transform lies on the positive side of (the cocircuit C with $\lambda_{\mathcal{S}}(C)=+1$ and vanishing at) b and on the negative side of c. In terms of the subdivisions, this implies that $E\setminus\{a,c\}\in S_{E\setminus\{c\}}$ but $E\setminus\{a,b\}\notin S_{E\setminus\{b\}}$.

Taking $B=E$ and $\sigma=E\setminus\{a,c\}$, Definition 5.9(ii) says that $E\setminus\{a,c\}$ lies in a cell σ of S. In the same way as before we can prove that $c\in\sigma$, so that either $E\setminus\{a\}\subseteq\sigma$. But then, Definition 5.9(i) taking $B=E$, $\tau=\sigma$ and $B'=E\setminus\{b\}$ implies that $\sigma\setminus\{b\}$ and hence either $E\setminus\{a,b\}$ of $E\setminus\{b\}$ is contained in a cell of $S_{E\setminus\{b\}}$. The second would imply b to have zero signs in the picture and the first would imply a lying in the negative side of b.

(3) Here we consider the two reorientation cases separately:

(3.a) In the picture of the left, $\{a,b,c\}$ is the support of a spanning positive circuit of $\mathcal{M}^*$, so that its complement is a simplicial facet of $\mathcal{M}$. Thus, there is a unique cell τ in S containing $E\setminus\{a,b,c\}$. But the picture would imply that $E\setminus\{a,b\}\in S_{E\setminus\{a\}}$ (b is in the negative side of a, and similarly $E\setminus\{b,c\}\in S_{E\setminus\{b\}}$

and $E \setminus \{a, c\} \in S_{E\setminus\{c\}}$. With this we conclude, respectively, $c \in \tau$, $a \in \tau$ and $b \in \tau$, using Definition 5.9(i). But then S would be the trivial subdivision, which is not the case.

(3.b) The picture tells us that $E \setminus \{a, c\}$ is a simplex in both $S_{E\setminus\{c\}}$ and $S_{E\setminus\{a\}}$, so that it is contained in a cell τ of S, by part (ii) of Definition 5.9 applied with $B' = E \setminus \{a, c\}$. Since neither $S_{E\setminus\{c\}}$ nor $S_{E\setminus\{a\}}$ is trivial, τ contains neither a nor c. Thus, $\tau = E \setminus \{a, c\}$ is a cell in S.

Part (i) of Definition 5.9 implies then that $E \setminus \{a, b, c\}$ is a face of a cell in $S_{E\setminus\{b\}}$. Since $S_{E\setminus\{b\}}$ is not trivial and since $E \setminus \{a, b, c\}$ does not span $\mathcal{M} \setminus \{b\}$, either $E \setminus \{a, b\}$ or $E \setminus \{b, c\}$ is in $S_{E\setminus\{b\}}$. This would imply one of a or c to be in the negative side of b in the picture. □

Proof of (a)⇔(c) in Theorem 5.10:

(c)⇒(a) This follows easily from Lemmas 5.11 and 5.12. If $\mathcal{S}$ is a family of subdivisions consistent with S, part 1 of Lemma 5.11 implies that $\mathcal{S}$ defines a circuit signature, which is the circuit signature of a lift by Lemmas 5.12 and 1.3. Then part 2 of Lemma 5.11 implies that S is the associated lifting subdivision.

(a)⇒(c) Let $\widehat{\mathcal{M}}$ be a non-cyclic lift of $\mathcal{M}$ inducing S. For every $B \subseteq E$ the restriction $\widehat{\mathcal{M}}(B)$ is a non-cyclic lift of $\mathcal{M}(B)$ and defines a subdivision S_B of $\mathcal{M}(B)$. Let us check that the family $\mathcal{S}$ of (lifting) subdivisions of the different restrictions of $\mathcal{M}$ is a consistent family of subdivisions.

If $B' \subseteq B$, each face of $\widehat{\mathcal{M}}(B \cup \widehat{p})$ not containing $\widehat{p}$ intersected with B' is a face of $\widehat{\mathcal{M}}(B' \cup \widehat{p})$. This proves the first condition of consistency.

Let $\sigma \subset B$ be a rank r simplex of $\mathcal{M}$ and suppose that for every $b \in B\setminus\sigma$ we have σ contained in a cell of $S_{\sigma\cup\{b\}}$. This implies that, in the restriction $\widehat{\mathcal{M}}(\sigma \cup \{b, \widehat{p}\})$, σ lies in a facet not containing $\widehat{p}$. In other words, in $\widehat{\mathcal{M}}(B \cup \widehat{p})$, every element $b \in B \setminus \sigma$ lies either on the hyperplane spanned by σ or in the same side as $\widehat{p}$ of this hyperplane. Hence, σ lies in a facet of $\widehat{\mathcal{M}}(B \cup \widehat{p})$ not containing $\widehat{p}$. This proves the second condition of consistency. □

In [**30**], Theorem 5.10 is used to prove that all triangulations of cyclic polytopes are lifting. For doing this, the following two weaker reformulations of the definition of consistent family of subdivisions are given. Although in that paper only the acyclic realizable case is considered, the proofs of the two results apply to the general case without change.

LEMMA 5.13 (Rambau-Santos). *Conditions (i) and (ii) in the definition of a consistent collection of subdivisions are equivalent to:*

- (i') *For every cell $\tau \in S_B$ and for every $b \in B$ the set $\tau \setminus \{b\}$ is a face of a cell of $S_{B\setminus\{b\}}$.*
- (ii') *For every rank r simplex σ of $\mathcal{M}(B)$, if σ is contained in cells of both $S_{B\setminus\{b\}}$ and $S_{B\setminus\{c\}}$ for some pair of elements $b, c \in B \setminus \sigma$ with $b \neq c$, then σ is contained in a cell of S_B as well.* □

LEMMA 5.14 (Rambau-Santos). *If $\mathcal{M}$ is uniform, then condition* (i') *of Lemma 5.13 is equivalent to the following one:*

- (i") *For every cell $\tau \in S_B$ and for every $b \in B$, if $\tau \setminus \{b\}$ is spanning then it is a cell of $S_{B\setminus\{b\}}$.* □

In the case of triangulations, the definition of consistency can be simplified further. In the following statement we use the notation $\mathcal{P}(T)$ and $T|_A$ introduced before Proposition 5.3.

COROLLARY 5.15. *Let T be a triangulation of an oriented matroid $\mathcal{M}$ on a set E. Then,*

- *T is lifting if and only if there is a family $\mathcal{S} = \{T(B) : B \subseteq E\}$ of triangulations of the restrictions of $\mathcal{M}$ consistent with T.*
- *This is equivalent to saying that $T(E) = T$ and that for every $B \subseteq E$, $T(B)$ is a triangulation of the restriction $\mathcal{M}(B)$ satisfying:*
  (i) $\mathcal{P}(T(B))|_{B\setminus\{b\}} \subset \mathcal{P}(T(B \setminus \{b\}))$ *for every* $b \in B$.
  (ii) $T(B \setminus \{b\}) \cap T(B \setminus \{c\}) \subset T(B)$ *for some* $b, c \in B$.

PROOF. If T is a lifting triangulation, in the proof of Theorem 5.10 we can assume that the lift $\widehat{\mathcal{M}}$ inducing it is generic, meaning by this that the dual extension is in general position. In these conditions the consistent family of subdivisions obtained contains only triangulations. This, with Theorem 5.10, proves the first part. The second part is a straightforward rephrasing of Lemma 5.13 for the case of triangulations. □

Bibliography

[1] L. ANDERSON, Topology of combinatorial differential manifolds, *Topology* **38** (1999), 197–221.

[2] L. ANDERSON, Matroid bundles, in *New Perspectives in Algebraic Combinatorics*, (L. Billera, A. Björner, C. Greene, R. E. Simion, and R. Stanley, eds.) MSRI Book Series **38**, Cambridge University Press, 1999, pp. 1–21.

[3] M. AZAOLA, The Baues conjecture in corank 3, preprint 1999, to appear in *Topology*.

[4] M. M. BAYER, Equidecomposable and weakly neighborly polytopes, *Israel J. Math.*, **81** (1993), 301–320.

[5] M. M. BAYER AND B. STURMFELS, Lawrence Polytopes, *Canadian J. Math.*, **42** (1990), 62–79.

[6] L. J. BILLERA, P. FILLIMAN AND B. STURMFELS, Constructions and complexity of secondary polytopes, *Adv. Math.*, **83** (1990), 155–179.

[7] L. J. BILLERA, I. M. GEL'FAND AND B. STURMFELS, Duality and minors of secondary polyhedra, *J. Combin. Theory, Ser. B*, **57** (1993), 258–268.

[8] L. J. BILLERA, M. M. KAPRANOV AND B. STURMFELS, Cellular Strings on Polytopes, *Proc. Amer. Math. Soc.*, **122**(2) (1994), 549–555.

[9] L. J. BILLERA AND B. S. MUNSON, Triangulations of oriented matroids and convex polytopes, *SIAM J. Algebraic Discrete Methods*, **5** (1984), 515–525.

[10] L. J. BILLERA AND B. S. MUNSON, Polarity and inner products in oriented matroids, *European J. Combin.*, **5** (1984), 293–308.

[11] A. BJÖRNER, M. LAS VERGNAS, B. STURMFELS, N. WHITE AND G. M. ZIEGLER, *Oriented Matroids*, Cambridge University Press, Cambridge, 1992.

[12] J. BOKOWSKI AND H. ROHLFS, *On a mutation problem of oriented matroids*, preprint 2000. `http://www.mathematik.tu-darmstadt.de/~bokowski/mutations_submitted.ps.gz`

[13] J. A. DE LOERA, Nonregular triangulations of products of simplices, *Discrete Comput. Geom.*, **15** (1996), 253–264.

[14] J. A. DE LOERA, S. HOŞTEN, F. SANTOS AND B. STURMFELS, The polytope of all triangulations of a point configuration, *Doc. Math. J. DMV*, **1** (1996), 103–119.

[15] J. A. DE LOERA, F. SANTOS AND J. URRUTIA, On the number of geometric bistellar neighbors of a triangulation, *Discrete Comput. Geom.*, **21** (1999), 131–142.

[16] P. H. EDELMAN AND V. REINER, Visibility complexes and the Baues problem in the plane, *Discrete Comput. Geom.* , **20** (1998), 35–59.

[17] J. EDMONDS AND K. FUKUDA, *Oriented matroid programming*, Ph. D. Thesis of K. Fukuda, University of Waterloo, 1982.

[18] J. EDMONDS AND A. MANDEL, *Topology of oriented matroids*, Ph. D. Thesis of A. Mandel, University of Waterloo, 1982.

[19] I. M. GEL'FAND, M. M. KAPRANOV AND A. V. ZELEVINSKY, *Discriminants, Resultants and Multidimensional Determinants*, Birkhäuser, Boston, 1994.

[20] I. M. GEL'FAND AND R. D. MACPHERSON, A combinatorial formula for the Pontrjagin classes, *Bull. Amer. Math. Soc.*, **26** (1992), 304–309.

[21] B. HUBER, J. RAMBAU AND F. SANTOS, The Cayley Trick, lifting subdivisions and the Bohne-Dress theorem on zonotopal tilings, *J. Eur. Math. Soc. (JEMS)*, **2**:2 (2000), 179–198.

[22] M. LAS VERGNAS, Extensions ponctuelles d'une géométrie combinatoire orienté, in: *Problémes combinatoires et théorie des graphes (Actes Coll. Orsay 1976)*, Colloques Internationaux, C.N.R.S., No. 260, pp. 265–270.

[23] C. L. LAWSON, Software for C^1-interpolation, in *Mathematical Software III* (John Rice ed.), Academic Press, New York, 1977.

[24] C. W. LEE, Regular triangulations of convex polytopes, in: *Applied Geometry and Discrete Mathematics-The Victor Klee Festschrift*, (P. Gritzmann and B. Sturmfels eds.), Dimacs Series in Discrete Math. and Theoretical Comp. Science 4, 1991, pp. 443–456.

[25] C.W. LEE, Subdivisions and triangulations of polytopes, in *Handbook of Discrete and Computational Geometry* (J.E. Goodman and J. O'Rourke eds.), CRC Press, New York, 1997, pp. 271–290.

[26] R. D. MACPHERSON, Combinatorial Differential Manifolds, in: *Topological Methods in Modern Mathematics: a Symposium in Honor of John Milnor's Sixtieth Birthday*, Publish or Perish, 1993.

[27] J. G. OXLEY, *Matroid Theory*, Oxford University Press, 1992.

[28] J. RAMBAU, Triangulations of cyclic polytopes and higher Bruhat orders, *Mathematika*, **44** (1997), 162–194.

[29] J. RAMBAU, *Circuit admissible triangulations of oriented matroids*, preprint, ZIB-Berlin, 2000. Available at `http://www.zib.de/PaperWeb/abstracts/ZR-00-45/`

[30] J. RAMBAU AND F. SANTOS, *The generalized Baues problem for cyclic polytopes I*, in "Combinatorics of Convex Polytopes" (K. Fukuda and G. M. Ziegler, eds.), *European J. Combin.*, **21** (2000), 65–83.

[31] J. RAMBAU AND G. M. ZIEGLER, Projections of polytopes and the Generalized Baues Conjecture, *Discrete Comput. Geom.*, **16** (1996), 215–237.

[32] V. REINER, The generalized Baues problem, in *New Perspectives in Algebraic Combinatorics*, (L. Billera, A. Björner, C. Greene, R. E. Simion, and R. Stanley, eds.) MSRI Book Series **38**, Cambridge University Press, 1999, pp. 293–336.

[33] J. RICHTER-GEBERT, Oriented Matroids with few mutations, *Discrete Comput. Geom.*, **10** (1993), 251–269.

[34] C. P. ROURKE AND B. J. SANDERSON, *Introduction to Piecewise-Linear Topology*, Springer-Verlag, 1982.

[35] F. SANTOS, On the refinements of a polyhedral subdivision, preprint 1999. Available at `http://www.matesco.unican.es/~santos/Articulos`

[36] F. SANTOS, A point configuration whose space of triangulations is disconnected, *J. Amer. Math. Soc.*, **13**:3 (2000), 611–637.

[37] E. SCHÖNHARDT, Über die Zerlegung von Dreieckspolyedern in Tetraeder, *Math. Ann.*, **98** (1928), 309–312.

[38] B. STURMFELS, *Gröbner bases and convex polytopes*, University Lecture Series **8**, American Mathematical Society, Providence, 1995.

[39] B. STURMFELS AND G. M. ZIEGLER, Extension spaces of oriented matroids, *Discrete Comput. Geom.*, **10** (1993), 23–45.

[40] G. M. ZIEGLER, *Lectures on Polytopes*, Springer-Verlag, New York, 1994.

Editorial Information

To be published in the *Memoirs*, a paper must be correct, new, nontrivial, and significant. Further, it must be well written and of interest to a substantial number of mathematicians. Piecemeal results, such as an inconclusive step toward an unproved major theorem or a minor variation on a known result, are in general not acceptable for publication. Papers appearing in *Memoirs* are generally longer than those appearing in *Transactions*, which shares the same editorial committee.

As of November 30, 2001, the backlog for this journal was approximately 6 volumes. This estimate is the result of dividing the number of manuscripts for this journal in the Providence office that have not yet gone to the printer on the above date by the average number of monographs per volume over the previous twelve months, reduced by the number of volumes published in four months (the time necessary for preparing a volume for the printer). (There are 6 volumes per year, each containing at least 4 numbers.)

A Consent to Publish and Copyright Agreement is required before a paper will be published in the *Memoirs*. After a paper is accepted for publication, the Providence office will send a Consent to Publish and Copyright Agreement to all authors of the paper. By submitting a paper to the *Memoirs*, authors certify that the results have not been submitted to nor are they under consideration for publication by another journal, conference proceedings, or similar publication.

Information for Authors

Memoirs are printed from camera copy fully prepared by the author. This means that the finished book will look exactly like the copy submitted.

The paper must contain a *descriptive title* and an *abstract* that summarizes the article in language suitable for workers in the general field (algebra, analysis, etc.). The *descriptive title* should be short, but informative; useless or vague phrases such as "some remarks about" or "concerning" should be avoided. The *abstract* should be at least one complete sentence, and at most 300 words. Included with the footnotes to the paper should be the 2000 *Mathematics Subject Classification* representing the primary and secondary subjects of the article. The classifications are accessible from `www.ams.org/msc/`. The list of classifications is also available in print starting with the 1999 annual index of *Mathematical Reviews*. The Mathematics Subject Classification footnote may be followed by a list of *key words and phrases* describing the subject matter of the article and taken from it. Journal abbreviations used in bibliographies are listed in the latest *Mathematical Reviews* annual index. The series abbreviations are also accessible from `www.ams.org/publications/`. To help in preparing and verifying references, the AMS offers MR Lookup, a Reference Tool for Linking, at `www.ams.org/mrlookup/`. When the manuscript is submitted, authors should supply the editor with electronic addresses if available. These will be printed after the postal address at the end of the article.

Electronically prepared manuscripts. The AMS encourages electronically prepared manuscripts, with a strong preference for $\mathcal{A}\mathcal{M}\mathcal{S}$-LaTeX. To this end, the Society has prepared $\mathcal{A}\mathcal{M}\mathcal{S}$-LaTeX author packages for each AMS publication. Author packages include instructions for preparing electronic manuscripts, the *AMS Author Handbook*, samples, and a style file that generates the particular design specifications of that publication series. Though $\mathcal{A}\mathcal{M}\mathcal{S}$-LaTeX is the highly preferred format of TeX, author packages are also available in $\mathcal{A}\mathcal{M}\mathcal{S}$-TeX.

Authors may retrieve an author package from e-MATH starting from `www.ams.org/tex/` or via FTP to `ftp.ams.org` (login as `anonymous`, enter username as password, and type `cd pub/author-info`). The *AMS Author Handbook* and the *Instruction Manual* are available in PDF format following the author packages link from `www.ams.org/tex/`. The author package can be obtained free of charge by sending email to `pub@ams.org` (Internet) or from the Publication Division, American Mathematical Society, P.O. Box 6248, Providence, RI 02940-6248. When requesting an author package, please specify AMS-LaTeX or AMS-TeX, Macintosh or IBM (3.5) format, and the publication in which your paper will appear. Please be sure to include your complete mailing address.

Sending electronic files. After acceptance, the source file(s) should be sent to the Providence office (this includes any TeX source file, any graphics files, and the DVI or PostScript file).

Before sending the source file, be sure you have proofread your paper carefully. The files you send must be the EXACT files used to generate the proof copy that was accepted for publication. For all publications, authors are required to send a printed copy of their paper, which exactly matches the copy approved for publication, along with any graphics that will appear in the paper.

TeX files may be submitted by email, FTP, or on diskette. The DVI file(s) and PostScript files should be submitted only by FTP or on diskette unless they are encoded properly to submit through email. (DVI files are binary and PostScript files tend to be very large.)

Electronically prepared manuscripts can be sent via email to `pub-submit@ams.org` (Internet). The subject line of the message should include the publication code to identify it as a Memoir. TeX source files, DVI files, and PostScript files can be transferred over the Internet by FTP to the Internet node `e-math.ams.org` (130.44.1.100).

Electronic graphics. Comprehensive instructions on preparing graphics are available at `www.ams.org/jourhtml/graphics.html`. A few of the major requirements are given here.

Submit files for graphics as EPS (Encapsulated PostScript) files. This includes graphics originated via a graphics application as well as scanned photographs or other computer-generated images. If this is not possible, TIFF files are acceptable as long as they can be opened in Adobe Photoshop or Illustrator. No matter what method was used to produce the graphic, it is necessary to provide a paper copy to the AMS.

Authors using graphics packages for the creation of electronic art should also avoid the use of any lines thinner than 0.5 points in width. Many graphics packages allow the user to specify a “hairline” for a very thin line. Hairlines often look acceptable when proofed on a typical laser printer. However, when produced on a high-resolution laser imagesetter, hairlines become nearly invisible and will be lost entirely in the final printing process.

Screens should be set to values between 15% and 85%. Screens which fall outside of this range are too light or too dark to print correctly. Variations of screens within a graphic should be no less than 10%.

Inquiries. Any inquiries concerning a paper that has been accepted for publication should be sent directly to the Electronic Prepress Department, American Mathematical Society, P. O. Box 6248, Providence, RI 02940-6248.

Editors

This journal is designed particularly for long research papers, normally at least 80 pages in length, and groups of cognate papers in pure and applied mathematics. Papers intended for publication in the *Memoirs* should be addressed to one of the following editors. In principle the Memoirs welcomes electronic submissions, and some of the editors, those whose names appear below with an asterisk (*), have indicated that they prefer them. However, editors reserve the right to request hard copies after papers have been submitted electronically. Authors are advised to make preliminary email inquiries to editors about whether they are likely to be able to handle submissions in a particular electronic form.

Selected Titles in This Series

(Continued from the front of this publication)

711 **V. G. Kac, C. Martinez, and E. Zelmanov,** Graded simple Jordan superalgebras of growth one, 2001

710 **Brian Marcus and Selim Tuncel,** Resolving Markov chains onto Bernoulli shifts via positive polynomials, 2001

709 **B. V. Rajarama Bhat,** Cocylces of CCR flows, 2001

708 **William M. Kantor and Ákos Seress,** Black box classical groups, 2001

707 **Henning Krause,** The spectrum of a module category, 2001

706 **Jonathan Brundan, Richard Dipper, and Alexander Kleshchev,** Quantum Linear groups and representations of $GL_n(\mathbb{F}_q)$, 2001

705 **I. Moerdijk and J. J. C. Vermeulen,** Proper maps of toposes, 2000

704 **Jeff Hooper, Victor Snaith, and Min van Tran,** The second Chinburg conjecture for quaternion fields, 2000

703 **Erik Guentner, Nigel Higson, and Jody Trout,** Equivariant E-theory for C^*-algebras, 2000

702 **Ilijas Farah,** Analytic guotients: Theory of liftings for quotients over analytic ideals on the integers, 2000

701 **Paul Selick and Jie Wu,** On natural coalgebra decompositions of tensor algebras and loop suspensions, 2000

700 **Vicente Cortés,** A new construction of homogeneous quaternionic manifolds and related geometric structures, 2000

699 **Alexander Fel′shtyn,** Dynamical zeta functions, Nielsen theory and Reidemeister torsion, 2000

698 **Andrew R. Kustin,** Complexes associated to two vectors and a rectangular matrix, 2000

697 **Deguang Han and David R. Larson,** Frames, bases and group representations, 2000

696 **Donald J. Estep, Mats G. Larson, and Roy D. Williams,** Estimating the error of numerical solutions of systems of reaction-diffusion equations, 2000

695 **Vitaly Bergelson and Randall McCutcheon,** An ergodic IP polynomial Szemerédi theorem, 2000

694 **Alberto Bressan, Graziano Crasta, and Benedetto Piccoli,** Well-posedness of the Cauchy problem for $n \times n$ systems of conservation laws, 2000

693 **Doug Pickrell,** Invariant measures for unitary groups associated to Kac-Moody Lie algebras, 2000

692 **Mara D. Neusel,** Inverse invariant theory and Steenrod operations, 2000

691 **Bruce Hughes and Stratos Prassidis,** Control and relaxation over the circle, 2000

690 **Robert Rumely, Chi Fong Lau, and Robert Varley,** Existence of the sectional capacity, 2000

689 **M. A. Dickmann and F. Miraglia,** Special groups: Boolean-theoretic methods in the theory of quadratic forms, 2000

688 **Piotr Hajłasz and Pekka Koskela,** Sobolev met Poincaré, 2000

687 **Guy David and Stephen Semmes,** Uniform rectifiability and quasiminimizing sets of arbitrary codimension, 2000

686 **L. Gaunce Lewis, Jr.,** Splitting theorems for certain equivariant spectra, 2000

685 **Jean-Luc Joly, Guy Metivier, and Jeffrey Rauch,** Caustics for dissipative semilinear oscillations, 2000